Global Warming
The Unsettled Science

Spike Hampson, PhD

Table of Contents

INTRODUCTION 1

THE SCIENCE 3

Bugs in the Theory 3
The Models Suck 5
A Misguided Assumption 7
Chasing Statistical Confidence 8
"That Damned, Elusive Pim . . . , *Tipping Point*" 11
The Temperature Record 12
The Infamous "Pause" 14
Why Lightweight Boxers Don't Fight Heavyweights 16
Sea Level Issues 17
What About the Earth's Internal Heat? 26
The Temperature Effect of Volcanic Eruptions 26
Now for the Humans 27
The Golden Dice 28
The Expected Disasters 29
Monkey Business 40
The Precautionary Principle 45

THE SCIENTIFIC CONCENSUS 49

"Settled Science" Does not Exist 49
"97% of All Climate Scientists Agree . . ." 51
The IPCC Juggernaut 53
Pseudoscience in a Tux 54
The Dark Side 55

HOW DID WE GET HERE? 57

INTRODUCTION

On January 20th of 2015, in his state of the union address, President Obama said: "And no challenge – no challenge – poses a greater threat to future generations than climate change." Such a categorical statement is not based on "settled science." Historical temperature data do not strongly support the theory and a large number of reputable scientists do not accept that increases in global temperature will threaten human existence. In both senses the science is far from settled.

Many argue that global warming is a technical issue that can only be resolved by trained scientists. Such experts can provide the solid factual foundation upon which societal judgments are made, but it is the general populace and their recognized leaders who must decide what course of action will best serve the general interest. For this more democratic process to properly occur, the ordinary citizen needs to understand the global warming theory well enough to ask sensible questions and to challenge unsupported assertions. In short, the average citizen must become better informed about the global warming idea.

Any ordinary individual with an active intellect can understand both the theory of global warming and the data that (fail to) support it. The opinion held by our president is not well-informed and my mission here is to explain in simple language why not. This is, in short, a skeptic's treatise.

Skepticism can serve the interests of science by identifying potential flaws in a commonly held view of how the world works. If a skeptic is challenging "settled" science it should be easy for mainstream scientists to counter the skeptic's criticisms with compelling logic – but only logic that is strongly supported by hard data.

These days, skepticism regarding the seriousness of global warming is not encouraged. Scientists, politicians, and the media all tend to view calamitous global warming in the same way that our president does: as an established truth that nobody should challenge. If global warming is a belief system, then one can regretfully understand that there might be an abhorrence of heterodoxy. If it is science, however, then there is no excuse for such an intolerant attitude regarding skepticism.

Skeptics do not reject the idea that the earth is warming. The greenhouse effect is real; warming does occur when CO_2 is emitted into the atmosphere; humans do contribute to those CO_2 emission. Nobody denies that these things are true; everybody agrees that they are. <u>The critical question is whether the magnitude of these events is sufficiently great that we need to be concerned.</u>

Global warming advocates are unwilling to accept that nothing be done about the fact that the earth is warming because they view the trend as a serious problem. This is why they feel threatened by the skeptics. But when the nuts and bolts of global warming theory are examined they don't assemble into a coherent whole. This is the conclusion of a skeptic who has looked at the science and found it unconvincing. What follows is an elaboration on the particular ways in which scientific data fail to give strong support to the theory.

THE SCIENCE

Bugs in the Theory

The prevailing theory holds that CO_2 emissions are causing dangerous increases in the average global air temperature. The physics are clear: atmospheric CO_2 does indeed capture heat. But the magnitude of this effect is not great and could never on its own account for the sharp rises in temperature that global warming believers like President Obama anticipate. In fact, a doubling of the CO_2 currently in the atmosphere would increase the ambient air temperature by less than two degrees Fahrenheit. If the amount of CO_2 in the atmosphere continues to increase at its current rate until the end of the century it will not double from its current level and will, therefore, increase ambient air temperature less than two degrees.

For CO_2 to increase the global temperature two more degrees Fahrenheit, the level of CO_2 would have to double again. The relationship is such that CO_2 levels must rise by ever greater amounts or else temperature will rise at a slower and slower pace.

Climate scientists recognize that CO_2 alone cannot cause troublesome rates of temperature increase but those who believe in the theory are convinced that associated feedback mechanisms will magnify this modest amount of heating caused by CO_2. In particular, the theory holds that warmer air absorbs more water vapor, which is much more effective at capturing heat than CO_2 is. As temperatures increase due to CO_2 in the atmosphere, the air will absorb more water vapor, and this is expected to greatly magnify the heating effect. It is the increase in water vapor – and not CO_2 – that is expected to drive atmospheric temperatures much higher.

The problem is that the expected amount of warming from additional water vapor simply has not materialized. As the air warms it should increase the amount of water vapor, which in turn should lead to even more warming. This may be happening but the magnitude of the effect is less than expected. Nobody knows why, but atmospheric temperature has not been rising at the pace forecasted by the global warming models.

Here is a possible explanation. Increasing levels of water vapor in the atmosphere probably result in greater amounts of cloud cover, and some types of clouds are very effective at reflecting solar radiation back out into space. In short, clouds have the potential to cause global cooling by deflecting some of the incoming solar radiation.

Because there is no good way of estimating future cloud cover and thus the quantity of solar radiation lost to space, the scientists who construct mathematical models for the purpose of predicting global warming have to make assumptions about it for their equations. Lacking data, they have little choice – but the assumptions are nothing but educated guesses and educated guesses are not good science.

The mathematical models used to forecast global temperature replicate a diverse set of atmospheric dynamics that are well understood, but, as the cloud cover example shows, not all relevant processes get included. If cloud cover and other poorly understood dynamics only have a marginal effect on global temperature then the models may succeed in accurately forecasting the future, but if the models fail to give accurate predictions then their designers have probably overlooked something important.

All the models used for forecasting average global temperature suffer from the same problem: they fail to accurately forecast future temperatures. We know this because the models have been around for two or three decades and so far they have predicted a lot more warming than has actually occurred. This suggests that the models are overlooking something important, and the fact that they all have been wrong in the same direction (more warming than subsequent data indicated) leads to the suspicion that they may all be overlooking the same thing.

A climate model is useful only if it accurately predicts the future. Even if it does predict the future well in the short run, there is still a possibility that its understanding of atmospheric dynamics is flawed and that the flaws will not be discovered until some later time when circumstances have changed. But if from the outset the model does not predict future temperatures accurately then it must be based on at least one faulty premise.

The climate models are of course incomplete: all models of any type simplify reality to some degree. This is acceptable as long as the model forecasts prove accurate. But if a model generates inaccurate forecasts then it would be foolish to rely on it when making decisions about the future. If a climate model incorporates the best scientific understanding about what causes global temperature to change, but fails to accurately predict changes in that global temperature, then the science is not settled.

The complexity of the physical world is so great that any model attempting to capture the dynamics of global temperature change must choose which natural processes are most likely to influence the temperature regime and must ignore all the other possibilities. To deal with those many natural processes judged to be of lesser concern, a common approach is to use existing historical data on temperature, run the model starting at a fixed point of time in the past, and then compare the output from the model with the actual temperature data for the subsequent years. To the extent that the model fails to properly capture that historical rate of temperature change, it can be mathematically adjusted so as to mirror as closely as possible the actual sequence of temperature changes that did occur. In this fashion, the scientist who likes to do models manages to combine a prediction based on a theoretical understanding of physics with a mathematical fudge factor that can be thought to include the net effect of all natural processes not explicitly incorporated into the model.

There may be a great many unknown dynamics influencing that fudge factor and it is impossible to know much about how they interact with each other. Nevertheless, the scientific investigator responsible for developing the model at least gets the illusory sense that the unknowns have been limited to a prescribed, secondary role in the model. It is an illusion because somewhere in that box of unknown dynamics there may be one of great causative importance – something that actually governs both CO_2 and temperature. Perhaps not: perhaps the model has captured and made explicit the most important source of causation, but the modeler cannot be confident about this until the unknown dynamics locked up inside that black box are ferreted out and the box itself has become vanishingly small. The best indication that the box has been diminished enough is when the model consistently makes very accurate predictions.

All the models used for forecasting average global temperature suffer from the same problem: <u>they fail to accurately forecast future temperatures</u>. We know this because the models have been around for two or three decades and so far they have predicted a lot more warming than has actually occurred. Why rely on models that have not proven themselves capable of accurate forecasts?

Some might argue that by correctly predicting the direction of temperature change (warming rather than cooling) they are at least exhibiting a certain fundamental level of success. But this is unpersuasive for two reasons. First, to make a prediction that in a random situation has a fifty percent chance of being right is not a significant accomplishment. Second, since global temperature has been fitfully warming ever since the end of the last ice age the prediction that such warming will continue is hardly impressive.

The Models Suck

Those who claim that the science of climate change is settled are contending that we now understand the physical and chemical processes that govern global temperature – understand them well enough, at any rate, to make accurate predictions about what will happen to global temperature whenever the complex system experiences a modification. This level of understanding is presented to the public in the form of temperature predictions generated by climate change models. Clearly, we should accept the notion of settled science only if the temperature predictions made by the models are proving to be accurate. But they are not.

Figure 1 uses a black line to show the actual change in global temperature from one year to the next and uses a gray zone to show the change predicted by the various global warming models that the International Panel on Climate Change has relied on in its most recent report. The model predictions are shown as a zone rather than a line because different models generate somewhat different results and because some latitude has to be given to the likelihood that reality will deviate from prediction due to nothing more than chance.

Figure 1

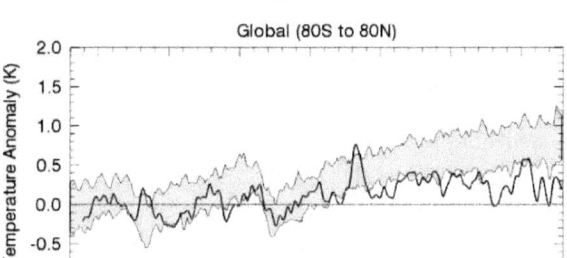

Temperature anomaly is the amount that temperature varies from the average for the time period 1979-1984. The thick, black line is the actual temperature record. The gray band is the confidence zone surrounding the average predicted temperature of all the IPCC models. After the year 2000, those models generally predict more warming than actually did happen.

If actual temperature falls within the gray zone, the prediction can claim to have been correct but if it does not then by its own admission the prediction has been wrong. Just before the year 2000, the model predictions shifted from being usually correct to usually wrong.

If the models no longer generate accurate predictions, how can the non-scientist have confidence that global warming science is "settled"? Lots of different explanations have been given for why the models were wrong – but the very appearance of such explanations undermines the proposition that the science is settled. Here is an example of an article that does nothing more than come up with possible explanations for why the models have been overestimating temperature increase:

http://www.nature.com/news/climate-change-the-case-of-the-missing-heat-1.14525

Since none of these explanations has yet achieved any reasonable level of scientific certainty, it is fair to conclude that the global warming science still is not settled.

Only by accepting the premise that unprecedented global warming will occur in the near future do we have any "evidence" that a catastrophe is looming. But a prediction of the future with no solid documentation that something similar has happened in the past requires extraordinarily compelling evidence of inevitable consequences from natural laws. Such evidence is not being provided. Instead, we are being told that the science is settled and the experts know best. The fact that all the respected models have made bad predictions is proof that they are failures and cannot in any way be considered as settled science.

The models are in trouble because the average global temperature has stopped rising very much. To a greater or lesser extent, all the reputable models have forecast a strong increase in the global temperature – not constantly upward on a year-by-year basis but reasonably so if we look at five year steps instead of one year steps. The problem is that the actual measured global temperature has hardly increased in the 21st century.

Some will argue that this is only a "pause," but to refer to this as a pause is based on the presumption that one knows what will happen in the future. Maybe it is a mere pause, but the use of that word implies a knowledge of the future that none of the models was able to predict. Since the models were unable to predict the "pause" in global warming, how reliable can they be?

The models have proven to be lousy at predicting global temperature rise but their defenders continue to support them, evidently based on the curious notion that year-to-year fluctuations in temperature are nothing but "noise" - a noise that somehow masks the fundamental long-term trend. But really, to call this noise is to suggest that science cannot be expected to explain short-term fluctuations in a temperature regime, even when those fluctuations often are greater than the total amount of temperature rise expected to occur over several decades.

The following open letter to Senators Inhofe, Rubio, and Cruz elaborates on why the models are failing to provide accurate predictions:

https://wattsupwiththat.com/2015/04/14/open-letter-to-u-s-senators-ted-cruz-james-inhofe-and-marco-rubio/

We are asked to believe that science is capable of thoroughly explaining the big picture but should not be expected to comprehend the little things. What would you think of a weather forecaster who told you that the atmospheric system is too complex to permit a good prediction of tomorrow's daytime high temperature but that his understanding of the big picture is so well developed that he can, with 95% confidence, assure you that the daytime high a month from now will be 68 degrees?

In addition to their proven inability to predict future global temperatures, the models have other serious weaknesses that, although less obvious, are in fact even more damaging to their reputations. The most profound of these has to do with how temperature increases vary with increasing distance above the ground.

The models expect temperature to increase faster in the upper atmosphere than at ground level. This is not simply a possibility that the models say may happen; it is a necessary consequence of the physical and chemical processes that underpin the global warming theory. But in fact we have temperature data for the past few decades that contradict this prediction. In recent decades the ground-level temperatures have warmed at a more rapid pace than temperatures in the upper atmosphere. This means that the mathematical formulas used by the models to evaluate global temperature change are based on a flawed understanding of atmospheric dynamics.

Not only do the models predict that temperature increases will be greater in the upper atmosphere than at ground level; they also predict that this upper level heating will occur more acutely in the tropics than anywhere else. But once again, the data contradict the theory: there is no indication that upper atmosphere heat is accumulating more rapidly in the tropics (see the final few paragraphs of this Wikipedia article):

https://en.wikipedia.org/wiki/UAH_satellite_temperature_dataset

Once again, the theory does not match the data.

The data, incidentally, are respected by climate scientists and come from two different sources: weather balloons and satellites. These two technologies are very different but they are in agreement that temperature rises have been faster at ground level than in the upper atmosphere. Their confirmation of each other increases the already high likelihood that the data are right and the models are wrong.

For the ordinary person, there is no obvious reason why temperature .increases should be expected to rise faster at higher levels in the atmosphere. Only those who have a detailed knowledge of the mechanical and chemical dynamics of global warming would naturally expect it. But even so, the ordinary person can readily see that when a physical condition predicted by all reputable model builders fails to materialize then there must be a fundamental misunderstanding in the "settled science."

A Misguided Assumption

There is another way in which all the reputable models may be failing to properly predict atmospheric conditions, and in this instance an untrained person can easily understand the nature of the problem. The models all compute that as temperature rises in the atmosphere there will be a diminishing rate at which the build-up of heat is dissipated by radiation out into space. There has been little data available regarding this issue, but of course it has a big influence on the amount of heat that gets retained in the atmosphere.

I am aware of one recent study in which measurements indicate the opposite – that heat build-up in the atmosphere leads to a higher rather than lower proportionate loss to outer space.

http://link.springer.com/article/10.1007%2Fs13143-011-0023-x

This is only one study, and to be fair there may be others that suggest the opposite. What the one study does do, however, is highlight the fact that the question may not be settled and that climate modelers may only be guessing at the rate of global heat loss. Anyone can see that accurate predictions of future global temperature are impossible if this dynamic is not clearly understood.

Simple logic suggests that the modelers' assumption regarding this matter will likely cause their models to forecast catastrophic rises in temperature sooner or later. By stipulating that the warmer the atmosphere gets the tougher it will be for that heat to escape into space, the modelers are building in a virtual guarantee that their forecasts for a warming earth will spiral towards an overheating disaster.

We do not know whether these commonly cited climate models yield up stable results over, say, a full millennium since they never are run to produce forecasts that far into the future. But if the models presume that increases in atmospheric temperature will slow the pace at which heat dissipates into space then they would appear to be inherently unstable. That lack of stability would work in both directions: warmer temperatures would cause a diminished rate of heat dissipation into space whereas colder temperatures would encourage an increased rate of heat dissipation into space. Each would lead to runaway temperature change, just in opposite directions. If the earth's heat balance really worked like this the atmosphere would have disappeared long ago.

One should think that any model of atmospheric dynamics would avoid incorporating an inherently unstable positive feedback loop since the actual history of the global atmosphere has shown reasonable stability for millions of years. Of course, if there is compelling evidence that extra atmospheric heat suppresses heat dissipation into space then this dynamic <u>should</u> be built into the models, but if this were the case it is inconceivable it would not have emerged as a major explanation for why we should expect an accelerated rate of atmospheric heating in the next few decades. This would be a big red flag that everybody would be talking about. Since it is not, the skeptic naturally questions whether this isn't an unfortunately misguided assumption inadvertently incorporated into the models.

In any event, we know that in deep geological time (in the time of the dinosaurs, for example) levels of CO_2 in the atmosphere were much, much greater than today – thousands of parts per million rather than the mere 400 parts per million we currently have. If the models of today were applied to conditions at that time, I should think they would likely have anticipated a far greater existential threat to (non-human) life on the planet – a threat that evidently failed to be realized. Of course, the overall composition of the atmosphere at that time might have been much different, thereby causing the models to project differently, but should not the climate scientists explain all this to us rather than just proclaiming everything to be "settled science?"

Chasing Statistical Confidence

There is another problem with the global warming models that rarely gets discussed. Built into all of them is a recognition that some aspects of atmospheric dynamics are not understood – are, in fact, responsible for a certain amount of unpredictable variation in temperature year-to-year. In short, these models are probabilistic rather than purely mechanical – largely because they do not have a perfect understanding of the phenomenon.

No two runs from a probabilistic model are exactly the same, but different runs of the model are usually close to being the same, with diminishing numbers of runs that deviate a lot from the average. It is standard practice to run a model many times and to then average together all the separate outputs by computing both the mean (average) value and the standard deviation (amount of deviation from the mean) for each separate year throughout the time span that the model was programmed to consider. The mean value for a given year becomes the actual temperature forecast and the standard deviation is used to establish a temperature range within which the actual temperature is expected to fall 95% of the time. The conventional, albeit arbitrary, presumption is that if the actual temperature ends up falling within that 95% zone then the model has predicted accurately whereas any actual temperature that falls outside the zone is taken as evidence that the model failed to correctly predict.

But here is what nobody talks about: a model with narrower zones of confidence yields more useful results than one with wider confidence zones. The narrower the confidence zones the more the model is able to say: "I have a thorough understanding of atmospheric dynamics and when I do multiple runs my year-by-year forecast testifies to this fact because my confidence zone is so skinny." But which of the two models has a zone of confidence within which the actual temperature is more likely to fall? The one that is less useful. The vaguer model with broader confidence levels will often "outperform" the more robust model, but only because it is less rigorous in its science. Beware of fat confidence zones: not only are they less useful as a predictive tool; they also indicate a weaker understanding of the atmospheric dynamics.

To give an abstract example, let us imagine two different meteorological forecasts. Suppose that one forecaster says that tomorrow's high temperature will be 73° or 74° whereas a competing forecaster tells his viewers that tomorrow's high will be between 75° and 85°. When it turns out that the actual high on the following day is 75°, the first forecast ends up being considered wrong whereas the second gets judged as right. The "rightness" of that second forecaster almost certainly is not because of a better understanding of atmospheric dynamics; it is the consequence of a less demanding standard of performance.

Now let us consider the nature of the global temperature forecasts used in the set of Intergovernmental Panel on Climate Change (IPCC) reports published in 2013. Figure 2 appears in Chapter 9 of the Volume entitled <u>Climate Change 2013: The Physical Science Basis</u>. It shows the actual temperature record (the thick, black line), the projected temperatures generated by 36 different climate models (the many thin lines), and the

Figure 2

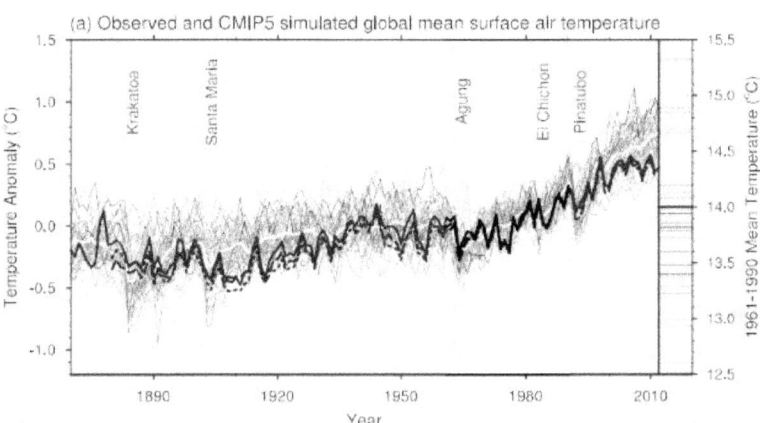

Temperature anomaly refers to how much the temperature for a given year differs from the average temperature for the period 1961-1990. Major volcanic eruptions occurred at the time their names appear in the chart. The thin graph lines in different colors are average projections for individual CMIP models. The thick white line that runs through them is the average of all the thin lines. The black line is the actual temperature record.

When viewing the graph, remember that the models <u>only predict from the 1980's onward</u>. For the century before then all the models were simply using already known temperatures to refine their results. Also, most of the models "anticipate" the temperature drop associated with the Mount Pinatubo eruption – a strong indication that they were adjusted to fit the real data even into the 1990's.

computed average temperature projected for all those different climate models (the thick, white line). Please remember that the white line is the average of all the thin lines and notice the recent trajectory of the white line (which shows forecast temperature) relative to the black line (which shows actual temperature).

Until almost 1980, these climate models did not exist so for the first 110 years shown in the graph the models were not actually doing any predicting. Instead, that earlier time period was a baseline of already known temperatures to which each model could fit its results before projecting out into the unknown future. Of course, different models were initially developed in different years, so some of those models did not come on line until significantly after 1980. In addition, some of the earlier models were able to use subsequent years to readjust their starting points so as to better fit reality.

When all this forecasting gained momentum in the early 1980's, the white and black lines initially followed similar trajectories but then in the late 1990's they began to diverge. Since then, the average forecast for the models has anticipated ever-warmer temperatures while the actual temperature regime has hardly changed at all.

Now let's look at the IPCC forecast of global temperature for the remainder of the twenty-first century. Figure 3, taken from the IPCC Executive Summary (published in 2013), gives temperature forecasts under two scenarios, one of which is depicted in light gray and presumes that insignificant efforts are made to abate CO_2 emissions and the other of which is shown in dark gray and indicates what might be accomplished if very strong abatement actions are taken.

Figure 3

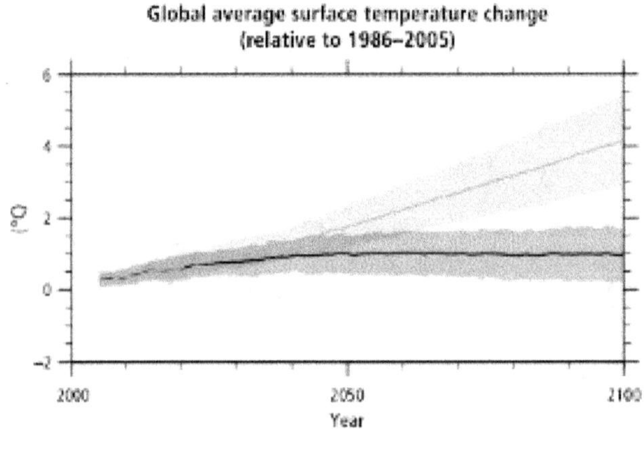

Global average surface temperature change
(relative to 1986–2005)

Let us first consider the light gray, business-as-usual scenario. By the year 2100, the IPCC expects the global temperature to have increased by 4.2 degrees Centigrade over the arbitrarily selected baseline average temperature of the 1986-2005 period. But notice the breadth of the light gray zone flanking the predictive line through its middle. The zone is what I would call "fat." The IPCC graph implies that if in the year 2100 the global temperature is anywhere between 2.8 and 5.9 degrees above the baseline then their forecast will have been "correct." But this is an awfully big zone, is it not? The confidence zone is 3.1 degrees wide when the actual predicted temperature increase is only 4.2 degrees.

Notice, incidentally, that even if humanity is able to follow the "best case scenario" depicted in dark gray, the maximum level for its confidence zone in the year 2050 is 1.4 degrees above the baseline. The recent Paris Accord specified that if global temperature is allowed to increase by only 1.1 degrees in the next few decades (by 2050) then we will "most likely" pass a tipping point after which global temperature spirals out of control. And yet, the IPCC doesn't even address the fact that we might do everything possible to curtail CO_2 emissions and yet still pass the tipping point. The IPCC is remarkably explicit about what humanity must do in the coming years (stop emitting CO_2) but is evidently unconcerned about the fact that even our best efforts may be for naught. Such broad confidence zones – graphically depicted but not accorded any significance in the accompanying narrative – should make us suspicious about the accuracy and the utility of the models.

In fact, these confidence zones shouldn't even be used to judge what the actual temperature is expected to do since their proper function would be to tell us what some totally new, similarly-vetted climate model would likely predict. In other words, the IPCC is misusing the concept of confidence zones. The IPCC appears not to care much about whether forecasts of future temperature are accurate or useful; the main objective seems to be graphic reinforcement of dire temperature predictions.

"That Damned, Elusive Pim . . . , *Tipping Point"*

It used to be that global warming was a concern because escalating temperature levels might modify the biosphere so much that humans would have trouble adapting. This vague but intuitively persuasive contention was compelling back in the early days when the forecasts for temperature increase were suggesting that the amount of heating could be around 10-15 degrees Fahrenheit over the span of a mere lifetime. Ten to fifteen degrees in a single lifetime! That was something people could sink their teeth into! That was something even people with a weak imagination might find worrisome.

But now, with the passage of a few decades during which global warming advocates have moderated their claims regarding the pace of heating, the specter of catastrophic change has become rather less compelling. The most recent IPCC report, for example, forecasts a seven degree rise in temperature over the next 85 years. This is still a lot of heating but it is an awful lot less than 12.5 degrees. Not only that, it is – as has already been discussed – a rate of estimated warming that significantly exceeds the actual rate that temperature records tell us has occurred so far in the 21st century. The recent unwillingness of actual temperatures to rise as much as the forecasts expected has created a sort of existential threat for the global warming advocates who – although still firm in their belief – find it harder and harder to convince others to join the movement.

When the forecast for temperature rise by the year 2100 is seven degrees while the actual temperature trend for the first fifteen years of the century suggests it might end up being only one degree – well, the urgency of the supposed threat becomes even less compelling because people lose confidence in the prognosticators. It was time for the global warming advocates to turn to a new line of reasoning.

A few years ago, the idea of a tipping point was nowhere to be seen, but now it is at the heart of the global warming argument. It shifts the emphasis away from the total amount of warming to be expected over the next, say, 100 years and instead claims that a magical temperature level exists beyond which everything will spin out of control. A set of unspecified positive feedback loops will kick in and force temperature higher no matter what we do to stop it. We must, therefore, take action immediately and keep temperature from crossing that fateful threshold. Global warming advocates cannot specify the actual tipping point temperature – but seem to be quite convinced that we are very close to it.

Tipping points certainly exist and have been delineated in science – both in terms of their quantitative parameters and in terms of their physical consequences. To give but one simple example, a block of ice is capable of existing whenever the water temperature is less than 32 degrees Fahrenheit, but if its mass is heated to anything more than that the ice melts and becomes water. There are many, many such tipping points in the natural world but the atmospheric temperature tipping point is quite unusual in that nobody knows the temperature at which it is activated or the irreversible physical processes that will ensue. Well, perhaps some scientists do know but little effort is made to inform the general public.

The recent Paris Accord on Climate Change codified a plan of action that was based entirely on the magical global warming tipping point. The final document paid very little attention to what atmospheric conditions might be like in the year 2100 and no attention to what might be done to cope with temperature change. Instead, it instituted a plan of action designed to insure that global atmospheric temperature not increase by more than two degrees. Why two degrees? Because it was presumed that a tipping point lurks in that general vicinity. Is this really science? I think not. The following URL reveals the extent to which an undefined tipping point is used to get everybody worried:

> http://www.independent.co.uk/environment/global-warming-passing-the-tipping-point-466187.html

The Temperature Record

Some will argue that regardless of whether the models are accurate, we must take seriously the temperature rise that actually has occurred in the past century. We do indeed have temperature records running back that far for a number of locations around the world and when averaged together they do clearly show an increase in the global temperature.

There are all sorts of weaknesses in this temperature data that might lead to measurement error. Just to name a few: the growth of cities creates urban heat islands that cause an artificial warming trend in the very locations where temperature records are most likely to be kept; weather stations come and go, making the selection of sites to use for calculating an average global temperature a somewhat subjective exercise; weather stations often have to be moved, which can cause sharp shifts in their recorded temperatures; devices and techniques used to take temperature measurements vary from place to place and change with time. In addition to these sources of potential error there is the alarming fact that the weather stations used to estimate global temperature are intensely concentrated in a few relatively small areas – especially the United States and Western Europe. Usable weather stations are very scanty in all the other continental areas; and over the oceans, which account for 70% of the earth's surface, they are virtually non-existent.

These are merely examples of ways in which the temperature data may not be reliable, but I mention them only to show that the global temperature record is most likely flawed. For the sake of discussion, I will assume that the temperature data are not flawed, but in reality the temperature record ought to be examined much more carefully than has been done so far.

But one takeaway that really should not be ignored is that the nature of the data and its awkward geographical distribution make it clear that any number purporting to be the global temperature should always be viewed as an <u>estimate</u>. We do not know the average temperature of the world; we are estimating it and our estimate is bound to include an unknown amount of error. When animated discussions revolve around whether or not a particular year was the hottest on record they are silly arguments because they ignore the fact that we really do not know the global temperature. Our computed numbers are just good-faith estimates.

But let us assume that the temperature record is very accurate. Figure 4 is NOAA's estimate of global temperature (change) since 1880. In the 115 years since 1900, the computed global average temperature has increased by about 1.6 degrees Fahrenheit. If one simply extrapolated this as a linear trend into the future, it would suggest that by the year 2100 the additional increase might be one more degree Fahrenheit. The models that have sounded the global warming alarm have projected much, much greater temperature increases for the next 85 years, so it is

incumbent on them to clarify why we should expect such an escalation in the pace of global warming. The IPCC graph, for example, gave a forecast of temperature increase by the end of the century of almost 7 degrees Fahrenheit (0.5 degrees Centigrade in 2015 to 4.2 degrees Centigrade in 2100 equals 3.7 degrees of change, which converts to 6.6 degrees Fahrenheit). Why should we believe that temperature is going to increase nearly 7 degrees when the current trend suggests an increase of only one degree? What is the specific dynamic that is expected to cause this acceleration and why is it not yet having any effect?

Figure 4
Global Annual Departure of Temperature From Average, 1880 - 2014

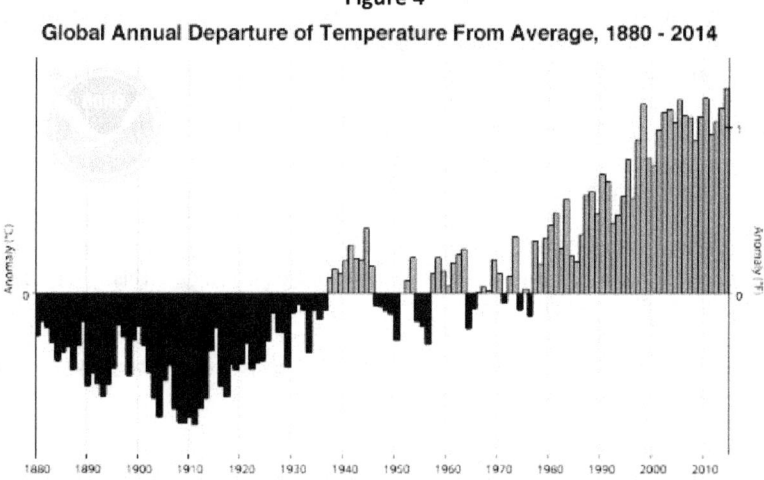

In the early 1990s most of the models were predicting even greater increases in global temperature over the course of the 21st century. Since then, modifications and adjustments to the models have scaled back the expected temperature increase, but as Figure 3 showed, the IPCC expectation for the year 2100 is still nearly seven degrees warmer than now. Why should we expect so much warming when (1) an extrapolation of the existing trend would only raise the temperature one degree and (2) the real 21st century trend is already lagging behind the model projections? If the answer were CO_2 emissions, then surely the models would be more accurate in their forecasts.

Some might argue that the recent warming trend has actually been exponential rather than linear, but so far there is no graphical indication of such a pattern. In fact, the global temperature record does not even show steady increase. From 1880 to about 1910 average global temperature actually declined, but then increased quickly in the 1910-1945 period. After that, from 1945 to about 1975, temperature drifted downward before reversing course and sharply rising until late in the 1990's. Now in the 21st century the average global temperature has stabilized and shown very little warming. This fitful pattern is pretty hard to square with any sort of exponential growth mechanism.

For the century between 1880 and 1980, the climate modelers knew the answer in advance and yet still failed to construct models that closely followed the decadal variations in temperature change that did occur during that time period (please refer back to Figure 1). Modelers will argue that unanticipated temperature swings probably were the result of periodic events like el niños, and this may indeed be the case. But considering that the models claim not only to capture all the important processes relevant to global temperature change <u>but also to incorporate a mathematical adjustment for all those supposedly unimportant ones</u>, it is disturbing that model results do not better reflect the actual pattern of temperature change during the 1880-1980 period. Any after-the-fact explanation for why they don't smacks of rationalization and subjectivity.

If the accumulation of CO_2 in the atmosphere is causing global warming, one would expect its pattern of increase to govern that of temperature increase. But it does not.

Every year, as regular as clockwork, the atmospheric CO_2 level increases by two or three parts per million. The annual growth in global temperature, on the other hand, is unpredictable: some years up and some years down, sometimes a big change from the preceding year and sometimes not. Processes that we don't completely understand are causing this and the modelers are unable to account for it. They contend that their understanding of atmospheric dynamics is sufficient to permit confident predictions about global temperature rise over the next century, but their inability to model the fluctuations in temperature from year to year, and indeed from decade to decade, undermines their credibility. If they can't tell us what the temperature situation is going to be in twenty years (and their poor record for the past twenty years proves this) then why should we accept that they are able to predict what the situation will be in nearly 100 years?

The global warming theory has been with us for no more than four decades and yet for only the first two of them did average global temperature increase substantially; the last two decades have not seen a significant amount of warming. This recent period of temperature stability has become so long that scientists who believe in global warming are having to propose explanations for why the warming trend has stopped. This is rationalization; this is after-the-fact justification. When you are wrong, the appropriate course of action in virtually all instances is to admit it.

The Infamous "Pause"

The most widely posited theory for the recent "pause" in global warming contends that CO_2-generated heat is being absorbed into the oceans. This is physically unlikely and totally unproven – and no credible theory has emerged to explain what would cause it. Ask yourself this: If atmospheric heat is being absorbed by the oceans, why did the process suddenly kick in about two decades ago? Why did it not operate before then?

Figure 5 shows both CO_2 content and global temperature year by year from 1959 to the present. Both lines do indeed trend upward but the upward slope of each is measured in different units that have been arbitrarily scaled by the constructor of the graph to give the illusory sense that the two phenomena are undergoing comparable rates of growth. The roughly parallel nature of those lines is meaningless.

The important thing is that the two lines show no similarity in shape. If the two lines both showed a constant rate of growth or – better yet – showed comparable patterns of change in that rate, then we would have good circumstantial evidence that the two processes are somehow linked. But they do not do this and thus fail to even imply any relationship. The two lines get from 1959 to 2015 by following totally different paths. There is no circumstantial evidence that one caused the other.

Their differences do not destroy the global warming theory since lots of other things besides CO_2 concentration may be influencing global temperature, but the extent to which the two lines differ is a good indication that those other things are important in the general explanation of why global temperature changes with time. And this is where the current global warming theory is most misguided: it contends that CO_2 is far and away the most important manipulator of average global temperature. The fact that these two lines are so different in shape means that <u>there must be</u> other things – big things; overlooked things – that contribute to average global temperature.

Figure 5

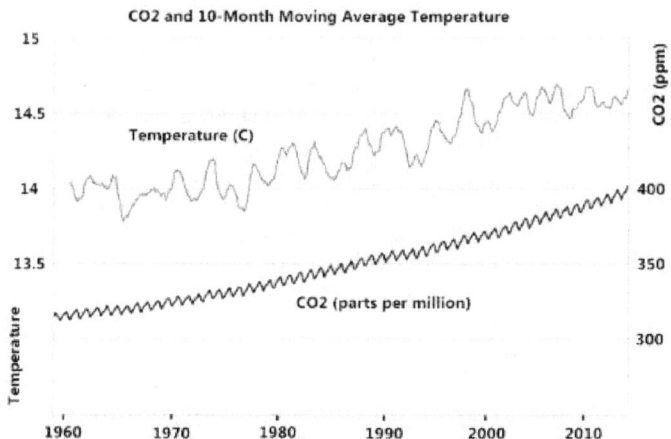

Since the two scales measure totally different things, the fact that the two variables have about the same slope means nothing. Also, the placement of temperature above CO2 is meaningless.

Besides, advocates of the idea that CO_2 concentration causes warmer temperatures need to confront an uncomfortable prehistoric reality. For great long sweeps of time, upward and downward shifts in levels of CO_2 and temperature did indeed closely match, but with temperature changing direction hundreds of years before CO_2 did. This is what the raw data say, although climate change advocates have been looking for ways to explain away the inconsistency. The following URL elaborates on the issue:

http://spectrum.ieee.org/energywise/energy/environment/carbon-dioxide-and-temperature-levels-are-more-tightly-linked

If changes in CO_2 concentration cause temperature changes, wouldn't the CO_2 have to change first? Climate change apologists discount the damaging prehistoric data by theorizing that the air bubbles in ice, from which CO_2 levels are measured, actually migrate upward in the ice to younger layers, thereby leaving the impression that they are younger than their actual age. But data are lacking that would confirm the idea.

If the prehistoric record says anything about causality, it is that temperature changes cause changes in the CO_2 level, and not the other way around. This purely circumstantial evidence is easy to reinforce with a persuasive theory: we know that the world ocean is a vast sink for CO_2 but that the warmer it gets the less CO_2 it can retain. Researchers claim that the oceans are warming. If this is true then their forced transpiration of CO_2 might increase the atmospheric concentration of CO_2. The ocean, after all, is a big player: it accounts for over 40% of all CO_2 emitted into the atmosphere.

This notion that oceanic CO_2 emissions are causing the rise in atmospheric CO_2 levels is just as theoretical as the global warming contention that CO_2 in the atmosphere drives temperature increase. Both rely on relatively simple concepts that struggle when confronted with real-world conditions – an abundance of poorly understood natural dynamics that bear on the question. Neither is settled science.

In the case of oceanic CO_2, for example, recent studies attempting to identify how much human activity affects it have run into all sorts of complications:

http://earthobservatory.nasa.gov/Features/OceanCarbon/

Simple explanations for why the ocean contains a certain amount of CO_2 look less likely to materialize today than they did a few decades ago. If global warming apologists are honest with themselves they will have to admit that the same thing is happening in their efforts to understand CO_2 in the atmosphere.

A big problem with the idea that the pause has been caused by absorption of heat into the ocean is that nobody can explain the mechanism. The ocean surface waters have not shown sufficient temperature increase in recent years so advocates for the idea are speculating that the heat is being absorbed by the deep ocean, for which we lack historical temperature records that might give any sort of confirmation. But really, if this is what is happening then why have the surface temperatures not also shown a significant rise? How can that atmospheric heat manage to bypass the ocean surface and only affect the deeps?

It is generally accepted within the study of climate that the atmosphere does not get heated much by the sun's rays directly. Instead, that solar radiation is absorbed at the earth's surface and then gets re- radiated back up into the atmosphere where (because of its longer wave lengths) it more effectively heats up the atmospheric molecules. In other words, climatologists believe that the temperature of the atmosphere is most governed by the amount of heat residing in the surface that is under it.

It is also a fundamental climatological principle that land heats and cools faster than water. Since the atmosphere adopts a temperature regime that reflects the underlying surface, this means that the air temperature over continents fluctuates much more than the air temperature over oceans – fluctuates more not just between day and night but also between summer and winter. Temperature regimes in continental interiors experience much greater swings than they do on small islands in the middle of an ocean. The amount of heat required to raise the temperature of a fixed volume of ocean water one degree is greater than it is for the same volume of continental earth, but both require hundreds of times more heat than is needed to do the same thing to atmosphere. When it comes to influence, the ocean is king, land masses are royalty, and the atmosphere a mere pauper.

Why Lightweight Boxers Don't Fight Heavyweights

Climate scientists are trying to convince us that atmospheric temperatures are heating up the oceans, but this completely contradicts the traditional understanding about what heats what. The density, the mass, of ocean water dictates that it will much more effectively alter the temperature of its proximate atmosphere than that insubstantial atmosphere can ever hope to heat the ocean. The density of water, after all, is about 800 times greater than that of air.

Consider two different objects that both have a uniform temperature of 70 degrees: a cubic foot of ocean water and a cubic foot of atmosphere. Which do you suppose contains the greatest absolute amount of heat? Obviously, the water does, by far. What this means is that water can heat air much more easily than air can heat water. Or to put it another way, water can heat atmosphere an awful lot faster that atmosphere can heat water.

Let us examine it in a slightly different way. We still use the two parcels of heated material – a cubic foot of atmosphere and a cubic foot of water – but we assume that there is a 10-degree temperature difference between them. When we place them in contact with each other, the warmer one will give up heat to the cooler one, until both parcels have the same temperature. The evening out of temperature will result in a balance that is much, much closer to the original starting temperature of the water than that of the air. Once the new equilibrium is established between the two parcels the air parcel temperature will have changed more than 9.9 degrees and the water parcel less than 0.1 degree.

Once we clearly understand this relationship between mass and embedded heat, it is hard to accept the global warming proposition that the oceans of the world are heating up (and therefore expanding, to create higher sea levels) because of their contact with an atmosphere that has increased its temperature by less than one degree in the last century. It just doesn't add up.

Sea Level Issues

Measurements regarding the melting of ice caps and the rise of sea level expose the degree to which global warming theory is not settled. In particular, the IPCC endorses an estimate for how much ice is melting in Antarctica and Greenland that is inconsistent with the estimate it endorses for how much sea level is rising. When the IPCC fails to present a logically coherent description of the science of sea level rise it is natural for any thinking person to question whether we really understand the phenomenon.

There is strong evidence that sea level rose a great deal at the end of the last ice age, and the idea that it was caused by melting ice is highly compelling. Today, however, sea level rise is at a very slow pace while direct measurement of global ice melt fails to account for it. Contemporary measurement systems claim the ability to detect outlandishly small amounts of both sea level change and global ice melt, but the results of their measurements do not confirm each other's accuracy. Should they not if we are to accept that the science is settled?

The climate change logic is that a warming atmosphere will lead to the melting of ice caps (Antarctica and Greenland) and the expansion of water in the ocean. These are the only mechanisms whereby global warming might contribute significantly to sea level rise. Of the two, only the melting of ice caps can raise sea level enough – quickly enough – to pose a serious challenge to human adaptability.

<u>A Serious Inconsistency</u>

The consensus that has been reached by climate science regarding the relationship between atmospheric warming and sea level rise is contained in a recent IPPC report entitled <u>Climate Change 2013: The Physical Science Basis</u>. A PDF of the report is available at:

https://www.ipcc.ch/pdf/assessment-report/ar5/wg1/WGIAR5_SPM_brochure_en.pdf

On page 7, it states that for the period 2002-2011 the average rate of ice loss from the Greenland ice sheet was 215 billion metric tons per year and from the Antarctic ice sheet was 147 billion metric tons per year. These are relatively small quantities. When the two figures are added together and converted to volume, and then distributed over the surface of the world ocean, they jointly contribute about one millimeter per year to sea level rise.

Two pages later, on page 9, the report contends that between 1993 and 2010 the sea was rising at a rate of 3.2 millimeters per year. In other words, the IPCC claims that sea level is rising about three times faster than ice cap melt suggests it should be. This is a rather glaring inconsistency for which the IPCC offers no explanation.

Quantifying the Heat Expansion of Sea Water

Of course, there does still remain the matter of sea water expansion associated with warmer temperature, but this is a well-understood phenomenon that can be expressed mathematically. The coefficient for the thermal expansion of water is 0.00021 (proportional increase in water volume associated with a one-degree Centigrade rise in temperature):

http://www.engineeringtoolbox.com/cubical-expansion-coefficients-d_1262.html

The rate of expansion varies depending on the initial temperature of the water, but the average surface temperature of the world ocean is so stable that the 0.00021 figure is sufficiently accurate for our purposes.

Referring once again to the same IPCC report, page 6 states that, ". . . ocean warming is largest near the surface, and the upper 75 meters warmed by 0.11 degrees Centigrade per decade over the period 1971 to 2010." Warming of 0.11 degrees Centigrade per decade means 0.011 degrees Centigrade per year.

When we multiply the IPCC estimate of annual warming for the upper 75 meters of the ocean (0.011 degrees Centigrade) by the coefficient of thermal expansion (0.00021), the result is a very small figure: 0.0000023. In other words, if we arbitrarily set the total volume of all ocean water in the 0-75 meter range at one unit, the IPCC is telling us that at the end of one year of ocean warming we can expect that volume to have increased to 1.0000023.

Since all ocean water expansion must necessarily translate into rising sea level, we can approximate the amount of that rise. But first, for comparative purposes, let us convert those 75 meters to millimeters: 75 x 1,000 = 75,000. Now let us apply the coefficient of expansion: 75,000 x 1.0000023 = 75,000.1725. In other words, based on the thermal expansion of that water in the top 75 meters of the ocean we can expect an annual sea level rise of 0.1725 millimeters. (Please note that this procedure yields a high estimate of sea level rise because it assumes that coastal zones of the world ocean are nowhere shallower than 75 meters and that a rising sea level will not increase the areal extent of the world ocean.)

Some will argue that the warming of the ocean extends much deeper than 75 meters, but a temperature profile of the ocean depths shows that below 300 meters, the temperature decline is precipitous. It is generally believed that oceanic surface waters do not mix with abyssal waters, except over very long time scales, and so additional climate-related heating of the ocean is bound to be limited to the surface area.

But how deep is that surface area? Perhaps surface heating does extend farther down than 75 meters, but the thermocline depicted in Figure 6 strongly suggests that any significant heating is limited to no more than about 300 meters of depth. Even if we presume that thorough surface mixing of ocean waters extends down to 300 meters, this would only quadruple the estimated amount of sea level rise – from 0.1725 millimeters to 0.69 millimeters.

Of course, the great bulk of ocean water is deeper than 300 meters, and this leaves open the possibility that mixing between surface and abyssal waters is greater than we realize and that a very slight amount of heat expansion throughout that enormous volume of deeper waters may be contributing an unknown but significant amount to sea level rise.

This, however, is highly improbable because the coefficient of expansion for water diminishes in colder water, until reaching zero at about 4 degrees Centigrade. In ocean waters that cold, a rise in temperature would not cause any expansion at all. The thermocline in the graph shows that all ocean waters deeper than about 1000 meters have temperatures that are very close to that 4 degrees Centigrade and thus would experience virtually no expansion if subjected to a slight amount of heating. Since most of the ocean is deeper than 1000 meters, most of the ocean <u>cannot</u> contribute to sea level rise. Only water that is substantially warmer than 4 degrees Centigrade has the potential to do such a thing.

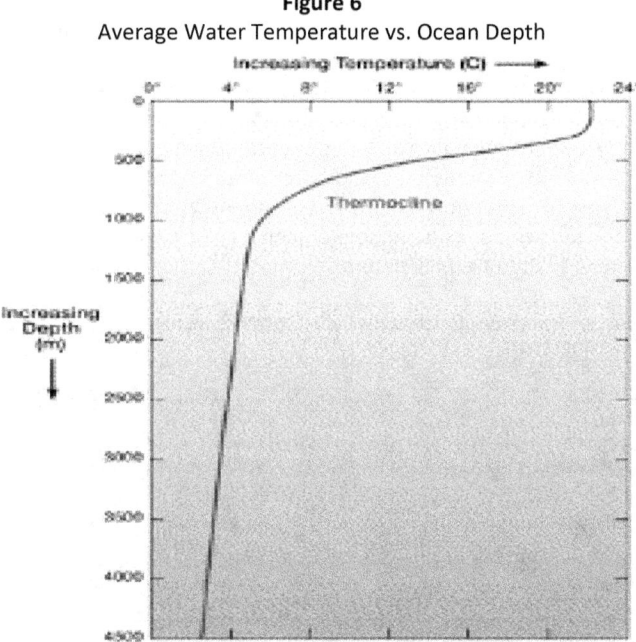

Figure 6
Average Water Temperature vs. Ocean Depth

The IPCC report goes on to say that there was likely to have been some warming of the ocean down to as much as 2000 meters, but does not specify its magnitude. The fact that the report specifies a numerical measurement of warming for the top 75 meters of the sea but offers no specifics for any water deeper than that suggests a lack of certainty regarding warming at greater depths – and in any event, for measurement of sea water expansion we now know that water deeper than about 750 meters would not expand significantly when warmed.

It is reasonable to conclude that if the IPCC estimate for ocean warming at the surface is correct then the expansion of sea water due to warming must lie somewhere between 0.1725 and 0.69 millimeters per year. When this is combined with the IPCC estimate for sea level rise due to ice melt (one millimeter per year), it becomes clear that climate change can only account for no more than 1.7 millimeters of sea level rise per year. This is about half of the 3.2 millimeters of sea level rise that the IPCC claims is happening each year. What accounts for the other half and why does the IPCC not address the question?

Either the global warming theory regarding the dynamics of sea level rise is wrong or the estimate of ice cap melt is wrong or the accepted measurement of sea level rise is wrong. Something is wrong somewhere. The science is not settled.

The extremely small coefficient of expansion and the physical inability of abyssal sea water to expand with warmer temperature strongly suggest that expansion of sea water due to warmer temperatures is too limited to increase sea level so much or so fast as to pose a threat to coastal communities around the world. Even if one uses a high estimate for sea water expansion (0.69 millimeters per year), for example, it would only account for less than 3 inches of sea level rise by the year 2100.

Are the Ice Caps Really Melting?

Sea ice (diminishing in the Arctic; increasing in the Antarctic) cannot affect sea level. Only the melting of land-based ice can raise sea level and for all practical purposes this means the ice caps of Antarctica and Greenland. Since the volume of Antarctic ice is ten times greater than Greenland ice, the glacial dynamics over Antarctica are more critical to the sea level question than the glacial dynamics over Greenland.

A recent NASA study claims that land-based Antarctic ice is increasing by about 83 billion tons per year. That is equivalent to 75 billion metric tons. One billion metric tons of melted water converts to one cubic kilometer, and a cubic kilometer of water would raise the global sea level by 0.00278 millimeters.

> https://www.nasa.gov/feature/goddard/nasa-study-mass-gains-of-antarctic-ice-sheet-greater-than-losses
>
> https://climatesanity.wordpress.com/conversion-factors-for-ice-and-water-mass-and-volume/

If sea level were affected by nothing else, this would cause an annual decline in sea level of 0.2 millimeter.

Wikipedia suggests that Greenland is losing about 195 cubic kilometers of ice per year, and that converts to slightly less than 0.55 millimeter of sea level rise.

> https://en.wikipedia.org/wiki/Greenland_ice_sheet#The_melting_ice_sheet

When the sea level effect of ice loss in Greenland is combined with that of ice gain in Antarctica, the net result is 0.35 millimeters of sea level rise per year.

If the NASA study is correct, it presents a challenge for global warming advocates. The following questions immediately arise:

> If the global atmosphere is warming, why is southern hemisphere ice cap ice actually increasing?
>
> Since the trend at present is that ice caps are melting at a rate that would only raise sea level about an eighth of an inch per decade, on what

scientific basis can it be contended that warmer temperatures will bring about a sea level rise of many feet by the end of the 21st century?

Since warmer global temperatures are not yet causing ice caps to melt significantly and since heat transfer from air to water would be too minor to significantly warm and expand the ocean, what mechanism is proposed by the global warming advocates to explain how atmospheric warming might cause a catastrophically rising sea level?

If the NASA study is correct, then then alarmists need to provide a convincing explanation for why global warming has led to ice accumulation in Antarctica. But really, even if this study is wrong, the annual amounts of ice that might be estimated to have melted in Antarctica would be much too small to contribute significantly to sea level rise. If melting ice is not the cause of sea level rise, what is?

The measured rate of sea level rise has been low: three millimeters per year translates to about an eighth of an inch per year. This rate has fluctuated from year to year but appears to be following a linear course and shows no obvious sign of recent acceleration. Figure 7, a NASA graph based on satellite data, confirms the lack of acceleration in sea level rise.

Figure 7

https://www.climate.gov/news-features/understanding-climate/2014-state-climate-sea-level

The vertical axis on this graph has been scaled so as to suggest that the sea level has been rising quickly, but the 80 millimeters that you see depicted on the vertical axis of the graph converts to only about three inches and the global trend line shown in the graph captures a sea level rise of about two and a half inches over a 22-year period. This is not enough sea level rise to justify the calamitous scenarios that global warming advocates describe when discussing the adverse consequences of sea level rise in the 21st century. An eighth of an inch per year, if continued throughout the rest of the century, would imply a sea level rise of only eleven inches between now and the year 2100. We would have 85 years to adjust to this.

Simple visual inspection of the trend line in the graph suggests that the rate of sea level rise has been steady and has not been accelerating. If global warming advocates are going to contend that the melting of ice caps in Antarctica and Greenland is accelerating the process of sea level rise, they need to bring some credible data to the table. Their own estimates of current sea level rise record a slow and evidently linear process. Why should skeptics believe that this pattern will evolve into a catastrophically rapid pace of change?

Sea level has been on the rise ever since the end of the last glacial period. To put the contemporary situation in context, the shape of the trend line in Figure 8 makes it clear that (1) sea level rise has been going on for millennia, (2) the period of truly rapid increase occurred between 15,000 and 7,000 years ago, and (3) the contemporary situation is in keeping with the languid pace of sea level rise that has prevailed since around 5000 BC.

The meltwater pulse that occurred between 15,000 and 14,000 years ago resulted in a sea level rise of about 30 meters. This computes out to an annual sea level rise of three centimeters. Some people contend that catastrophic global warming is on the verge of returning us to this rapid pace of sea level rise, but the current rate is only about one tenth as fast as that.

Estimates that contemporary sea level might rise as much as ten feet in the next century would require a return to that extraordinary rate estimated to have occurred 14,000-15,000 years ago, but such a thing simply cannot happen because the sea level rise in that earlier time period was surely the result of colossal melting of the very extensive ice sheets that covered much of North America and Europe. Now that those ice sheets have melted, there is not enough ice left to sustain such prodigious melting again. The melting of the glacial period ice sheets brought about a 130-meter sea level rise whereas the residual ice sheets of Greenland and Antarctica are calculated to contain only enough ice to cause about half that much additional sea level rise. It is not reasonable to think that glacial melt could again cause a ten-foot rise in sea level in a single century and even a glacial melt at, say, half that rate ought to be viewed as an improbable scenario that is purely theoretical in nature and that is totally unsupported by any contemporary trends in sea level rise or ice cap melt.

Figure 8

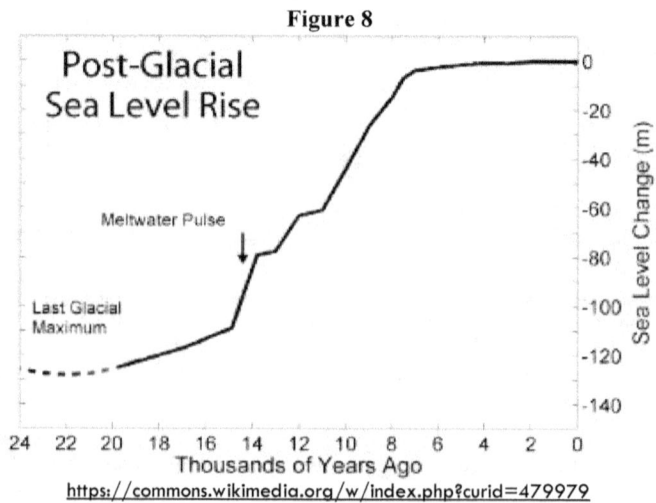

https://commons.wikimedia.org/w/index.php?curid=479979

Sea Level Rise is a Suspect Science

It is troubling to see so much significance being assigned to the miniscule amounts of sea level rise that have been detected in recent years. The contemporary figures for sea level rise may be accurate but I am doubtful: the confounding variables are so numerous and so disruptive that such precision seems highly improbable. The method of measurement is based on various assumptions that, although perhaps reasonable, mean the sea level figures are merely estimates. There is no way to confirm that the estimates are accurate.

Part of the problem is that we simply cannot measure actual sea level and must instead use statistical methods to estimate – relative to an unknown base – the amount of increase or decline that has occurred over a fixed period of time. The method for doing this has a reasonable theoretical basis: it presumes that multiple measurements of sea level taken regularly through time can be compared to each other as long as the measurement is always done using exactly the same tools and procedures. Even if there are flaws in this measurement system, they will always be the same flaws that, although capable of biasing the actual measurement results, will generally introduce the same bias in all instances, thereby leaving a residual measurement that can be considered as an accurate indicator of change. The theory is good but there are no benchmark data and there is no alternative methodology that can be used to cross check the accuracy of the sea level rise estimates.

What does exist, however is an array of measurement problems that could play havoc with the theory. Even assuming for a moment that the earth's crust is completely static and undergoes no changes over time, the world ocean alone introduces extraordinarily complex measurement problems.

The surface of the ocean is almost everywhere in a state of constant flux. At any given instant, passing waves are varying it by at least a few feet, passing swells are doing the same, and so are passing tidal flows. Waves, swells, and tides are at work everywhere on the world ocean, all the time. There is nothing predictable about the size of those waves and swells that constantly make landfall at any particular point along a coastline. Even tides, which do have a certain degree of predictability are not entirely so. It is therefore evident that even with perfect instrumentation, any single measurement of sea level taken at a given location at a particular point in time cannot be accepted as an accurate measurement of sea level. It is probably safe to say that the combination of waves, swells, and tides causes an unpredictable variation of sea level that amounts to at least a few feet. This "noise" is eliminated by taking multiple measurements and averaging them all together.

But statistical averaging obliges the investigator to establish that the variation of values around the mean is normally distributed. If the distribution is not normal, then conclusions drawn from the average (mean) figure are invalid. There is good reason to believe, however, that neither waves, nor swells, nor tides would return values that distribute normally. For the periodicity of each (from peak to peak of a passing wave, for example), much more of the time is spent near the extremes than in the transition. This suggests that the distribution for each would be the inverse of normal – a preponderance of readings near the extremes and a paucity of readings near the middle. Perhaps the net effect of all three would create a somewhat normal distribution, but I am doubtful.

It is not uncommon for scientists to "relax" the assumption of normalcy in the distribution of data, based on the notion that it will only compromise the final results a slight amount. The justification for this behavior may be reasonable in some instances but when the "noise" surrounding sea level change is measured in feet and the amount of sea level change being detected is measured in eighths of an inch, then I doubt that this is one of those instances.

As confounding as the undulating surface of the ocean may be, it is only a part of the larger problem. The larger problem is that the crust of the earth is not static. Both on land and under the sea, it is constantly moving up and down. It "floats" on an intensely hot soup of melted rock called the mantle that is believed to be fluid enough that its contents circulate in a convectional manner, much like the movement patterns of water in a pan that is heated from below.

Since the earth's crust is "afloat" on the mantle, it follows that any thickening of the crust in one locale will cause that area to settle deeper into the underlying mantle just as any thinning in a given area will result in upward movement.

But in fact no vertical movement of the crust can occur without a compensating movement elsewhere in the opposite direction. The compensation may be a smaller mass of material moving a larger distance in the opposite direction; it may be a greater mass of material moving a smaller distance in the opposite direction; it may even, on rare occasions, be the same amount of mass moving in the opposite direction the same distance. The point is that isostatic adjustment anywhere inevitably triggers a compensating reverse adjustment around its periphery that extends the isostatic activity a significant but unknown distance beyond the margins of the original area of interest.

Indeed, that adjustment around the periphery will cause a diminished but nonetheless inevitable adjustment in the reverse direction even farther out. The effect of an isostatic adjustment radiates outward much like waves generated by a stone dropped in a millpond. There, the generated waves are relatively large and close together near their source but quickly diminish in height and lengthen in periodicity with greater distance from the source. Isostatic adjustments of the earth's crust must do something similar. The vertical displacement of crustal waves quickly diminishes to very small quantities, but I don't think science has traced the wave pattern much beyond the original source.

When isostatic adjustment is viewed in this fashion, it leads to an image of the earth's crust not as a static and rigid topography upon which a sea level can be scribed but instead as a constantly undulating surface that confuses the very meaning of sea level by introducing isostasy as a pervasive manipulator of sea level virtually everywhere in the world. The complexities of isostatic adjustments are highlighted in an article in "Nature" magazine, an article that on the one hand presumes sea levels are rising catastrophically but on the other reveals how much vertical movements of the crust confound actual measurement of sea level change:

http://www.nature.com/news/climate-science-rising-tide-1.13749

We have, furthermore, the well-regarded theory of plate tectonics which contends that the crust of the earth consists of interlocking plates. These plates are believed to be moving horizontally, driven by the convectional currents in the mantle. The collision of adjacent plates, we believe, forces one of the plates to subduct under the other, thereby creating a thickening of the crust that inevitably triggers waves of isostatic adjustment.

It is crudely estimated that the lateral movement of plates often occurs on a scale of about an inch per year. This is eight times the annual amount of change we estimate in sea level rise. Of course, the relevance of crustal movements for sea level changes is tied to the amount of <u>vertical</u> shifting in plates. But if a plate is shifting an inch per year in a horizontal direction then the zone of compression between it and its neighbor may be causing a significant amount of vertical movement. A number of these subduction zones are located near continental coastlines. How do we know that they are not influencing measured sea level changes?

In addition to the isostatic adjustments that are very widespread around the world, there is growing evidence that the earth's crust is manipulated up and down by those convectional currents in the mantle – up wherever the current is ascending and down wherever it is descending. Perhaps as much as half of the upward movement of the crust in the vicinity of Hudson's Bay is now thought to be associated with convectional upwelling in the mantle rather than just crustal rebound from the disappearance of the last ice sheet:

http://science.howstuffworks.com/environmental/earth/geophysics/missing-gravity.htm

Furthermore, the enormous compression in a colliding plate might flex or distort it for some period of time. If, for example, compression were to flex a significant portion of some oceanic plates upward some small part of an inch per year it might be sufficient to account for a certain proportion of the measured annual sea level rise. Of course the flexion might be downward instead, but in either case plate tectonics could be altering sea level via a mechanism other than climate change.

All this may sound excessively theoretical, but in fact we do know that isostasy is not quantitatively insignificant. It is estimated, for example that isostatic rebound in northern Canada even today, many millennia after the disappearance of the glacial ice, is in the neighborhood of half an inch per year, which is four times the approximate rate of sea level change.

I am not trying to contend that these speculations about the vertical undulations of the earth's crust should be treated as established truths, but I would suggest that they are more probable than the assumption inherent in most climate change explanations for sea level rise that the earth's crust is – except for certain isolated zones of known isostatic adjustment – basically static and rigid. Up and down movements of the crust are most likely occurring virtually everywhere in the world.

On land, weathering and erosion are constantly thinning the crust in some areas while deposition is thickening it in others. Under the sea – and especially near coastlines where the question of sea level rise is most relevant – sedimentary deposition is greatly thickening (and therefore depressing) the crust. Both on land and under the sea, volcanism is thickening the crust (a force that we associate with volcanoes and lava flows but that probably is far more substantial as a creator of igneous intrusions that go largely undetected). Along zones where plates are moving away from each other, there might be thinning of the crust. Even if massive ice cap melting were to occur, the deposition of all that water (although only about a third as heavy as most rock material) would effectively thicken the crust. All these forms of globally distributed crustal change have the ability to alter sea level in any given locale and virtually no locale would be immune to the effects of at least one or two of them.

Even the underbody of the crust may be susceptible to thickening and thinning since the boundary zone between crust and mantle is most likely experiencing crustal melting or consolidation, depending on variations in the heat intensity of the mantle (probably hotter where the convection current is upwelling and somewhat "cooler" wherever the convectional current cycles downward).

We really do not know how much vertical movement of the crust is occurring, or at what speed, at most points on the earth's surface, but the forces at work probably cause some vertical movement of the crust most everywhere. We also have to recognize that these movements probably are not insignificant when measured relative to the estimated annual sea level rise of an eighth of an inch per year.

Under the circumstances, one must be skeptical about the accuracy of estimates regarding sea level change. The two technologies for collecting data regarding sea level are measurements coming from tide gauges and coming from satellites. Tide gauges are obviously plagued by severe limitations (including the inability to be read at so fine a scale as an eighth of an inch) and so satellite measurement of sea level is the only credible technology for achieving such precision.

Even though the satellite system of measuring sea level is extremely precise, the list of complications discussed above creates a high probability that those measurements are not accurate. Since reputable estimates for ice melt and sea water expansion indirectly imply a much lower rate of sea level rise than do the readings from satellites, one must decide which measurement system is wrong. Either that, or global warming advocates must come up with a climate-related explanation for sea level rise that goes beyond the notion that ice melt and water expansion are causing it.

But even if the satellite-based estimates of sea level rise are accurate, the rate at which the sea is currently rising does not justify the widely held view that coastal flooding will be pervasive by the end of the century. Such flooding will happen only if the ice caps start melting rapidly, but atmospheric temperature rise has not yet shown itself capable of initiating such a melt process. Direct measurements of ice cap ice contradict the theory that global warming is significantly melting the ice caps and the slow pace of sea level rise undermines the notion that coastlines will experience catastrophic submergence between now and the year 2100.

What about the Earth's Internal Heat?

If the oceans of the world are heating up, there are only two things that can be doing it: either solar energy or the earth's internal heat. Scientists have estimated the amount of internal heat that escapes through the earth's crust and have concluded that its contribution to ocean temperatures is relatively minor in comparison with solar radiation. But global warming advocates contend that solar radiation does not vary significantly in the short run (years or decades). If this is true, then short term heating of the ocean must come either from inside the earth or from heat in the atmosphere. Now the comparison is no longer between a major and a minor contributor to heating; it is a contestation between two minor ones, with a major one being held constant.

Once again we move into a realm of unsettled science. As you can well imagine, the estimates of oceanic warming from heat transfer through the crust are based on presumptions and assumptions, just as are estimates of oceanic warming from CO_2-based heat in the atmosphere. But let's play the odds here. How much heat can our atmosphere hold in comparison with the amount of heat contained inside the earth? Obviously, the internal source is a better bet. Lightweight boxers don't fight heavyweights.

The Temperature Effect of Volcanic Eruptions

Volcanic events, incidentally, are sometimes used by climate warming apologists as a partial explanation for why global temperature has recently failed to escalate as much as the models predicted. Their contention is that the particulate matter ejected by volcanoes tends to screen the earth from solar radiation. There have indeed been volcanic events in the past that were followed by a global chill lasting a year or two, and this does afford strong circumstantial evidence supporting the idea.

But wait: if particulate matter in the atmosphere prevents solar radiation from reaching the earth's surface it would have to either absorb that heat or reflect it out into space – and if it absorbed the heat then the atmosphere would be warming instead of cooling. It must be, therefore, that particulate matter manages to reflect that heat back out into space. We know that particulate matter encourages additional cloud cover so the most probable explanation is that the volcanic dust causes clouds which have highly reflective tops.

At the same time that volcanoes are injecting particulate matter into the atmosphere they are also introducing significant amounts of new heat. It does seem that this additional heat is more than offset by the screening effect of atmospheric particles, but the influence of volcanoes on temperature is clearly the net consequence conflicting processes. How do we know that the short term cooling actually documented when a volcano erupts is not offset by heat being added to the atmosphere – heat that may seem overshadowed by the temporary cooling but that could actually be a net addition to atmospheric heating in the longer term. The point is not that the climate warming apologists are wrong about the cooling effect of an eruption; the point is that they have reduced a complicated phenomenon to something that is unrealistically simple.

This is not a trivial issue. If additional cloudiness is not the explanation for why a volcano cools rather than heats the atmosphere, what is? If, on the other hand, cloudiness is the explanation for why a heating phenomenon actually causes cooling then surely the models of global warming are wrong to treat cloud cover as relatively unimportant in the equation that calculates global temperature change.

If in fact the particulate matter from a volcano somehow brings about atmospheric cooling directly rather than by encouraging greater cloud cover then the climate scientists are confronted with a new problem to solve: the amount of particulate matter in the atmosphere can cause significant variations in temperature and must be factored into their models (all atmosphere has particulate matter in it to a greater or lesser degree). The models already neglect the effects of variation in cloud cover; now they can be faulted for failing to consider variations in particulate matter as well.

Now for the Humans

Although the injection of CO_2 into the atmosphere does cause warming, only a very small part of all CO_2 emissions are the result of human activity. Widely accepted estimates are that over 95% comes from natural processes while only the small remainder comes from humans (see Figure 9 and then review this URL).

http://www.manhattan-institute.org/html/six-acres-and-deere-1455.html

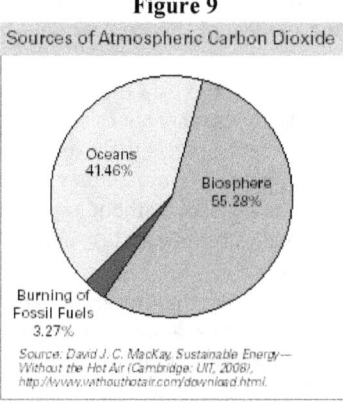

Figure 9

Global warming advocates do not argue with this. Instead, they claim that that small human contribution is an additional amount of CO_2 that upsets the natural balance between the uptake and release of the gas in the atmosphere. In this way, they are able to argue that humans rather than natural processes are driving the global warming phenomenon.

Ironically, this allows the skeptic to say with over 95% certainty that CO_2 buildup is caused by nature and not humans. The odds are better than 95 out of 100 that any randomly selected CO_2 molecule captured in the air had its origins from some sort of natural emission process. A theoretical sampling of CO_2 molecules taken from the atmosphere would lead in most instances to the remarkable conclusion that the human contribution to CO_2 build-up is not just small: it is statistically insignificant.

The Golden Dice

The amount of CO_2 in the atmosphere already is startlingly small – only about 400 parts per million (or one part in 2,500). The amount of CO_2 <u>added</u> to the atmosphere each year is far smaller yet: only two or three parts per million. Assuming humans are responsible for 5% of that, the math suggests that it takes humans 7-10 years to contribute only one part per million. This is a remarkably small quantity considering the mischief that global warming advocates claim for the process.

To more clearly visualize the volumetric significance of CO_2 in the atmosphere, imagine a jail cell that has no furniture and let its open space represent the atmosphere. Now toss in 400 dice that come to rest scattered around on the floor. Virtually all those dice are standard ivory, but scattered among them are a half dozen made of gold. The dice are the CO_2 in the atmosphere at present; the gold dice are the CO_2 that has been contributed by human burning of hydrocarbons. Once a year, the jailer opens the door and throws in two or three extra dice. That's the annual addition to the atmospheric stock of CO_2. But only by about the eighth year does the jailer's annual visit bring another gold die to add to the litter on the floor.

We can accept that that small scattering of dice on the floor of the cell is volumetrically sufficient to profoundly alter the characteristics of the atmosphere: it has already been established that a doubling of those 400 dice would raise the temperature in the cell by 1-2 degrees. If by the year 2100 there are indeed 800 dice on the floor (which would require much more than a simple maintenance of the warden's annual contribution) then, yes, the temperature can reasonably be expected to rise that small amount. It is logical, however, to reject the notion that "forced" additions of water vapor would greatly magnify the temperature increase since the models of this process have not been accurate in their predictions of such a process.

What simply cannot be accepted is the idea that the golden dice, which by the year 2100 would number approximately 20 – twenty out of eight hundred! – should be thought of as the principle culprit in the temperature rise (be it small or large). In fact, it makes no difference whether the dice are gold or ivory – they all have the same functional consequence. Just as craps might be shot without regard for the color of the dice, atmospheric dynamics will be equally altered by a CO_2 molecule regardless of whether it was introduced by humans or by nature.

Global warming advocates often claim that the anthropogenic sources of CO_2 are, in a sense, the last straw, a small but critical final addition to the atmospheric stock of CO_2 that manages to unbalance the natural system. If we are to believe this, then we must also believe that the more profound contributors to CO_2 in the atmosphere show very little change in their behavior over time. But that is almost certainly not true. We know that natural systems vary significantly decade-to-decade and century-to-century (decadal variations in global temperature before 1950 are a good example) and it is reasonable to think that natural CO_2 emissions would do the same. This means that the emotional appeal of the "last straw" argument is severely undermined since the natural system – and not humans – might provide it.

Global warming advocates contradict themselves when they contend that the reason global temperature has "paused" is that CO_2 suddenly is being absorbed into the world ocean. This would have to be a quick change in the absorption-respiration pattern of the ocean if it is to explain the temperature pause. By their own logic, the natural system has the potential to upset the balance of the natural system. We have already seen that the ocean accounts for about 40% of all the natural emissions of CO_2. Simple math tells us that if the oceanic respiration of CO_2 were to increase by only 10% (from roughly 40% to 44% of the total) then that would be as significant a contribution to CO_2 build-up as all human activity. Whatever the proportionate share of the CO_2 output the oceans represent, nature as a whole (including the oceans) would only have to vary by 3-4% to be as significant as a source of the gas as anthropogenic sources.

Of course, the warmist argument is that the human-caused CO_2 emissions are so important because they are in addition to the natural sources and because they always get added, every year. This is a logically coherent argument, but in recent years it has been outfitted with a corollary that weakens the original. Many experts are now claiming that if humanity fails to cut back significantly on its CO_2 emissions before the year 2025, then we will likely have passed a tipping point beyond which it will be too late to redeem the system: runaway global warming will be unavoidable.

Just ten short years from now. But this means that the same tipping point can be reached even if we stop all human-caused CO_2 emissions immediately. Why? Because the same end result would materialize if natural causes of CO_2 emissions were to have a single decade in which the total output was 3-4% above average. Given the usual variability of natural systems, and the customarily long time frames in which they function, a ten-year stretch of above average CO_2 emissions is quite possible.

But of course the tipping point is nothing more than a theoretical fabrication. Ask any global warming advocate to explain how it works or when it kicks in and I doubt you will get a convincing response.

The Expected Disasters

Let us assume that, in spite of my skepticism, rapid global warming is impending. Let us also grant that it is caused by CO_2 emissions into the atmosphere. Global warming advocates anticipate an array of catastrophic developments that need, they say, to be addressed immediately.

Of all the potential disasters that alarmists see coming, the most widely heralded is a rising sea level, but an earlier section of this essay has already exposed the lack of evidence in support of any catastrophic scenario. When data begins to show more substantial net losses of ice cap ice and quantitatively significant rises in sea level, then at that point humanity might seriously consider how to cope with a problem. But since the anticipated problem has not yet shown any sign of materializing it makes no sense to spend scarce resources attempting to solve a non-existent problem.

Other potential disasters that most distress the alarmists are (1) sea ice loss in the Arctic, (2) threats to reef ecosystems, (3) disruption of global agriculture, (4) readier spread of certain diseases, and (5) more extreme weather events. But are these real concerns? Probably not.

The Polar Bear Charade

Global warming is held responsible for a recent decline in Arctic sea ice, which in turn is blamed for a supposed decline in the polar bear population. This is misleading. Arctic sea ice did diminish significantly for a few decades, but not so much in the last seven years. Meanwhile, surveys indicate that the polar bear population has been growing rather than shrinking:

https://polarbearscience.com/2013/07/15/global-population-of-polar-bears-has-increased-by-2650-5700-since-2001/

The prospect of polar bear extinction is a suspect narrative driven by the media, evidently in an effort to convince people that sea ice melt in the arctic will have serious consequences for the environment. Before advancing the highly speculative notion that the polar bear population is being threatened by contracting sea ice, a reputable scientist would feel obliged to present solid evidence that bear numbers are actually dropping. Who is that scientist?

This alleged decline in the polar bear population is the "picture worth a thousand words" intended to show that rapid and accelerating sea ice melt in the Arctic is an ecological disaster. But is the Arctic ice actually disappearing? I would ask that a reasonable interpretation be made of what the chart in Figure 10 has to say. It was compiled by the National Snow and Ice Data Center.

Yes, there was a significant decline in Arctic sea ice in recent decades but did that really portend catastrophe? Does the chart really show an acceleration in the rate of sea ice disappearance? No, it does not. It shows significant contraction in the overall amount of sea ice, but not in the most recent years.

Figure 10

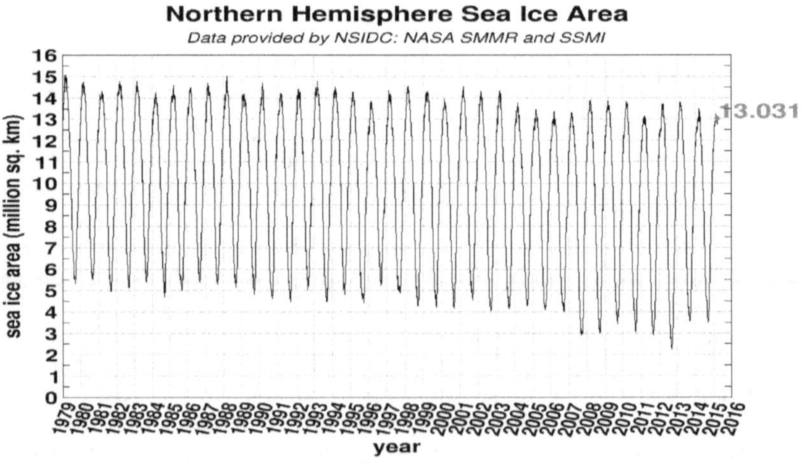

Northern Hemisphere Sea Ice Area
Data provided by NSIDC: NASA SMMR and SSMI

Besides, the argument is that a warmer global temperature is causing the polar ice to melt. But that makes no sense since Antarctic sea ice has been <u>increasing</u> in recent decades and the net global amount of sea ice has remained constant. If the global inventory of sea ice is stable, and thus not sensitive to a globally warming atmosphere, why should shrinkage of Arctic sea ice, representing about half of the global inventory, be taken as evidence that a rising global temperature is the cause? Failure to provide a good answer to this question should not allow the acceptability of a bad one. It would be better to say, "We don't know." And once one recognizes that there is contradictory evidence regarding the relationship between sea ice and average global temperature, it is natural to be skeptical about the idea that a great mass of sea ice in the Arctic is about to disappear and that a growing population of polar bears is on the verge of extinction. Would not settled science require that the above graph show an accelerating pace of Arctic sea ice melt? Would it not expect the polar bear population to be declining instead of growing?

Even if the extent of Arctic sea ice is sensitive to only a regional rather than global temperature regime, the theory of how the warming proceeds is challenged by the graph in Figure 10. Contraction of the Arctic sea ice should lead to ever increasing absorption of solar radiation into the surface because of the lower albedo of water and of dirty snow. This should drive Arctic temperatures up at an ever-increasing rate and should dissipate the Arctic sea ice ever more quickly. But the graph shows that the decline in Arctic sea ice has stabilized in recent years rather than spiraling downwards faster and faster. These data do not cooperate with the theory.

I do not think that the chart in Figure 10 makes a compelling case for the idea that Arctic sea ice is about to disappear, but even if it is, the consequences would not be entirely negative. The main concerns of the alarmists are that the disappearance of polar sea ice will drastically change the environment for the worse and will likely alter the flow pattern of major ocean currents. Let us consider the weaknesses in these speculative contentions.

We have been hearing for years that Arctic sea ice is declining rapidly, so what are the serious environmental issues that have begun to materialize as a consequence? All I have heard about so far is the plight of polar bears – and this myth has already been discredited. What other consequences have there been? Are there other species of plants or animals in decline? As far as I am aware, the Arctic continues to be a harsh and unforgiving environment in which a little extra warmth is unlikely to distress the caribou or challenge the tundra grasses. Frankly, the Arctic could use a little extra heat.

Speculation also runs riot that the warm and cold ocean currents will be affected as a consequence of sea ice melt in the Arctic. Concern is with what is called thermohaline circulation, a vertical mixing of surface and deep waters that is driven (in part) by temperature and density differences in the water. The idea is that sea ice melt – a source of fresh water - would inhibit the convectionally driven delivery of relative warmth into the high latitudes. The result would be . . . less heat in the Arctic! In other words, those who fret about such a thing are distressed by the prospect that warming in the Arctic could, by blocking thermohaline circulation, contribute to cooling in the Arctic.

Most of what one reads about this speculative possibility view it as potentially disruptive to Arctic environmental conditions in some way. But would it not be more sensible to think of it as a promising sign of inherent climatic stability? Since warming in the Arctic would encourage a physical process that tends to cool the Arctic, prospects for drastic environmental change there are diminished.

Although thermohaline circulation is a reality, the prime driver of ocean currents is prevailing wind patterns and tidal currents. The global wind patterns might be weakened if the Arctic were to get warmer, but probably would remain in place since the Arctic would still be relatively colder than all those latitudinal zones closer to the equator where heat is inevitably more abundant. Second, the flow pattern of ocean currents is largely determined by the positioning and configuration of the continental land masses that inevitably deflect those currents. Since the arrangement of the continents will not change much in the next few hundred years, the ocean currents are likely to remain where they are. As for the passage of currents through the Arctic region itself, the presence or absence of ice at the surface will not be determinative. Ice or no ice, the Gulf Stream will still flow over the top of Norway and across the Arctic Ocean, setting up cold counter-flows that exit via the Bering and Davis Straits.

I have also heard the argument that an ice-free and, warming polar region will likely melt most or all of the permafrost in the Arctic and sub-Arctic. This, it is contended, would be a bad thing because it would allow the escape of methane that is presently locked away in a frozen chamber. But permafrost is in fact the bane of the Arctic. It makes the construction of buildings and roads extremely expensive. It causes all rivers to be highly susceptible to extreme flooding every spring. It creates millions of square miles of boggy conditions throughout the tundra and taiga regions during the short summer season, thereby initiating an explosion of insect populations and greatly inhibiting overland travel by foot or dog team or indeed any form of off-road transit. In summary, permafrost more effectively discourages human settlement than do the extremely cold temperatures.

So obsessed are the global warming advocates with the evils of greenhouse gasses that they are unable to see the remarkable degree to which permafrost inhibits settlement. Either that, or they are using the issue as a lever to discourage the possibility of Arctic development. Even if methane emissions from the defrosted soil were accepted as a legitimate counter to the huge advantages that would accrue for human activity there, why is no consideration given to the enormous amount of additional biomass that would spring up in the Arctic when it becomes warmer. All that biomass would capture a substantial amount of CO_2 and yet no real effort is made to incorporate consideration of this inhibitive effect on atmospheric CO_2 levels if Arctic lands were to defrost.

Warmer temperatures in the Arctic would not be a totally bad thing. After all, Western Europe spent a few centuries trying to find a Northwest Passage through the region and their hope was for a purely seasonal opportunity to reach Asia and the Pacific quickly and easily. An ice-free polar sea would create a year-round maritime short-cut between the Atlantic and Pacific basins – a prospect to be deplored only if one's philosophy similarly regrets the epic construction of the Suez and Panama Canals.

If we look at the Arctic region today, we see a vast landscape that is very thinly populated. There is of course an aggressive environmental movement – supported by many – that wishes to keep it that way.

But this region is just another form of territorially based social injustice. Poor inner city neighborhoods are viewed as special zones in which the population needs access to better social conditions, but the poverty of south side Chicago, for example, is not more extreme than the rampant poverty to be seen in Kotzebue, or Yellowknife, or most any other Arctic settlement where social problems are widespread and the population generally consists of minority groups. This is just as true in Russia as it is in Canada and Alaska. If the people in the core of Detroit deserve better, then why should improved conditions be denied to those who live in the far north? Because the polar bears will die? Because we want it to stay cold up there? Because we think permafrost is, on balance, a good thing? It looks more like hypocrisy to me. But improving conditions in poor inner city neighborhoods is a form of economic development, is it not? More jobs, better schools, improved infrastructure, etc., etc. – these sorts of things are needed in the Arctic as well, and a willingness to develop them in inner cities but not in the far north is a pretty sorry commentary on the sincerity of social justice warriors. Bottom line: the environmental changes some expect to occur if the Arctic region warms will facilitate the sort of economic development that is thought to be needed in areas of high crime, dysfunctional families, high unemployment, low education, limited social and commercial facilities, and little available money.

It may very well be that negative consequences arising out of Arctic temperature increase and sea ice melt outweigh consideration for the people who live there, but global warming alarmists don't try to weigh the two sides of the equation; they stack the deck. They cherry pick the negative consequences on the environmental side of the equation and ignore the human side of the equation altogether. Arctic people are in fact as important as polar bears – and actually more numerous as well.

Now, finally, I would like to draw a picture of what might occur in the northlands if the Arctic Ocean were to become ice-free. It is an image that will set on edge the teeth of those who would like to set aside the Arctic as an untouchable preserve, but the people who actually live in the region would almost certainly welcome the change.

The open water would give much greater access to vast amounts of land – many millions of square miles of land – that heretofore have been almost inaccessible. This would be true throughout the Arctic regions that border on the frozen zones of the Arctic Ocean, but would be more transformative in the northlands of Russia and Canada than in the much smaller Alaskan sector. Both Russia and Canada have major rivers that flow north up into the Arctic Ocean and although these rivers might still freeze over in the winter months the severity of the freeze would be diminished by the presence of an open-water Arctic Ocean. Those rivers would be navigable to great distances inland. They would act as conduits for trade and commerce between the interior and the Arctic coast, with substantial settlements likely to emerge at or near the mouths of the rivers.

And these rivers are truly major. In Russia, the Ob, Yenisey, and Lena rivers are among the largest in the world and they would open up Siberia to development. Their river valleys would become north-south corridors of human penetration, linking the Arctic coast of Siberia to the tenuous thread of commerce and settlement arrayed along the trans-Siberian rail line far to the south. The trans-Siberian has struggled since the late 1800's to provide a basis for Siberian modernization. The enormous territory north of that line continues to this day to be a thinly populated homeland for a great variety of minority indigenous peoples, with outpost settlements of ethnic Russians only in a handful of resource-rich hot spots. With the permafrost gone and the temperatures a few degrees warmer, the summer season would extend noticeably and the boreal forests would move northward to displace much of the tundra zone. The longer growing season, a virtual certainty given the long days and short nights of summer, would permit the kinds of agriculture currently found much farther to the south: grain production, market gardening, and dairying.

In Canada the basis of the transformation would be much the same. The Mackenzie River (also one of the largest rivers) and the ice-free Hudson Bay would allow similar economic development along north-south linkages to the national heartland which at present is a relatively narrow band running along the country's southern perimeter. Canadians have long considered the north to be their frontier, but no longer would it be a far-flung network of isolated outposts dependent on bush pilots and radio signals for their connection to the larger world. It would become a region integrated into the rest of Canada by rail lines and highways and integrated power projects.

I think most global warming alarmists don't want any more development in the Arctic and thus deplore any kind of climate change in the region that might facilitate it. It is somewhat dishonest, however, to only identify negative effects of extra heat in the Arctic. It would be like proclaiming "Catastrophe!" if climate change were to cause more rainfall in a desert.

Is Dory Doomed?

Another global warming proposition is that absorption of CO_2 into the ocean is causing acidification of its surface waters. Much has been made of this idea in recent years and the stated concern is that it will kill coral reefs and shellfish (and, of course, poor Dory) because they rely heavily on the availability of calcium carbonate. This mineral diminishes when CO_2 causes ocean water to become more acidic. But there are serious problems with this theory.

First, we already know that the ocean accounts for roughly 40% of all CO_2 emissions into the atmosphere whereas human-caused CO_2 emissions only account for about 4%. However much humans might contribute to a reverse flow of CO_2 from the atmosphere into the ocean, it can never equal the amount of oceanic CO_2 being expired into the atmosphere. How can this slight drag on oceanic CO_2 emissions be held responsible for acidification? When we add in the fact that the warmer the ocean gets the more it gives up CO_2 to the atmosphere, the whole dynamic suggests that any human "contribution" to acidification will only diminish with time.

And then there is the awkward fact that by definition acidification of the ocean cannot be occurring. At present, the ocean is alkaline rather than acidic and diminution in its chemically basic character must be referred to as neutralization. Only when the pH of water drops below 7 (it is now at 8.1) will it be possible for acidification of the ocean to occur. Some might argue that I am playing with words here, but then simple justice would require that ocean acidification alarmists be accused of the same game (which they were the ones to start).

Rain water, incidentally, is acidic (a pH of about 5.6). This means that in the 70% of the world covered by oceans all the rain that falls is contributing directly to ocean "acidification." Rain that falls over land may or may not remain acidic as it flows to the sea (depending on the minerals it absorbs *en route*). If our natural system replenishes the alkaline ocean with acidic water, where is the hard evidence that the small and indirect contribution humans make to ocean "acidification" is somehow disruptive to the balance of nature? I am not arguing that humans are incapable of doing such a thing: I'm only asking that the contention be pinned down with more rigor before claiming it the right to be considered in any way definitive. Until then, it is only sensible to be skeptical.

None of this is meant to imply that coral reefs and shellfish are as robust and pervasive as ever. There is a great deal of evidence that both categories of calcium carbonate dependent life forms have undergone serious setbacks in recent decades. But there are already lots of culprits that have been identified as likely contributors to the problem, and many of them have human causes (siltification, chemical pollution, overharvesting, and impact damage, for example). To contend that oceanic absorption of CO_2 is also a big contributor is suspect since we have no documentation that the average global pH level of surface ocean water has dropped sufficiently to account for the declines, which tend to be local or regional in nature. Furthermore, there is no good way of identifying how much each of the different human-caused modifications to the marine environment is contributing to the overall picture. It is one thing to claim that CO_2 is a big part of the problem; it is another to prove it.

But as long as reef ecosystems and shellfish populations show significant variations in their relative health from one region of the world to another, it is problematic to claim that an overall trend towards neutralization of the surface ocean waters is the prime suspect in their demise. Geographic variability in the health of the populations cries out for local explanations rather than a global one.

Finally, since global warming advocates are increasingly drawn to the idea that the world ocean is being subjected to CO_2 induced warming, it is worth noting what the consequences would likely be for tropical reef systems. I do not accept that the oceans have eaten the missing heat that separates global warming projections of temperature from actually recorded temperature, but if in fact the explanation turns out to be true, then presumably the marginally warmer ocean waters will broaden the latitudinal perimeters of the tropical marine system such that coral reef will be able to expand into territory that was previously too cold. The very limitation of tropical reef to the tropics indicates that lack of heat in the water is at least as detrimental to reef health as too much heat. The CO_2 alarmists fail to point out this positive consequence of oceanic warming. Instead, they shift to a different argument about how CO_2 causes "acidification."

Disruption of Global Agriculture?

Global warming advocates fret that rising temperatures will modify the suitability of land for farming. There is every reason to believe that a longer growing season and warmer temperatures will indeed shift the geographic distribution of different types of farming, but there can be no denying that it will also open up enormous amounts of new land that until now have not been so suitable for farming. Canadians, Russians, and northern Europeans may very well become concerned if quantitatively significant sea level rise ever materializes, but I suspect they will not feel oppressed by the idea that they may have to adapt their farming practices to warmer temperatures.

As for the other extreme - the tropics – there is no good evidence that additional heat will seriously undermine agricultural practices there. After all, if we hold precipitation constant, we find that warmer temperatures generally permit the production of greater amounts of agricultural biomass – and I am aware of no evidence that this becomes untrue in tropical zones where the growing season is year-round. Indeed, agricultural productivity can often be very high even in the extremely hot desert environment as long as a way can be found to maintain an ample water to the plants.

Of course, we cannot hold precipitation constant since changes in temperature regimes will likely alter patterns of rainfall. But scientists cannot accurately predict this highly complex interaction. Since they cannot predict it, they really should refrain from making categorical statements about how rising temperatures will negatively affect farming.

About all we can say is that things will change. Whether the change will help or hinder agricultural production is not known. We do know, however, that when precipitation levels are held constant, higher temperatures will result in a greater and more diverse biomass on the landscape. And quantity of biomass is a pretty good indicator of how productive land might be for agriculture.

The only way I can visualize a decline in the innate potential of the globe for agricultural productivity is if global warming were to cause an expansion of arid and semi-arid realms. Recent evidence, however, suggests that the global extent of world deserts is shrinking and not expanding:

http://news.nationalgeographic.com/news/2009/07/090731-green-sahara.html

http://news.bbc.co.uk/2/hi/africa/8150415.stm

https://www.newscientist.com/article/dn2811-africas-deserts-are-in-spectacular-retreat/#.VSxHfvnF-ow

Farmers are not so unsophisticated that they cannot adapt to warming conditions by shifting over to cultivars that are more suited to heat. For the farmer, cold and dryness are the big enemies; heat is generally welcomed. Not only that, farmers have shown themselves to be adept at such adaptations, often switching from one cultivar to another in order to take advantage of such things as higher prices, less water consumption, and lower perishability of the crop. Adaptation is natural to the farmer who accepts the need to adjust to what climate has to offer instead of trying to change the climate to suit the crop.

In the last century the world has undergone very rapid urbanization such that today urban dwellers are much more numerous than farmers and hunter-gatherers and fishermen. Most of the experts in a field like climate science probably come from an urban background where an intimacy with the ways of the natural world is not possible. The result, I suspect, is hubris regarding their book learning and a serious deficiency of direct, experiential knowledge of nature – a lack of intuitive recognition of the degree to which natural forces govern our affairs. The farmer may be less deluded in this respect.

But, alas, it is more than mere ignorance. For global warming advocates to claim that an elevating global temperature will damage global agriculture, they must focus all their attention on possibly negative consequences and avoid consideration of possibly positive ones. How else is it possible to explain their contrarian view regarding what makes plants grow?

The red flag that most starkly reveals the warmist's lack of objectivity regarding the expected consequences of global warming for agriculture is the total avoidance of all discussion regarding the effects of CO_2 on plants. It is scientifically well-established that higher concentrations of atmospheric CO_2 stimulate plant growth and yet no mention ever is made of this good news:

http://www.climatecentral.org/news/study-finds-plant-growth-surges-as-co2-levels-rise-16094

http://www.nature.com/scitable/knowledge/library/effects-of-rising-atmospheric-concentrations-of-carbon-13254108

Although a more nuanced understanding of the relationship between atmospheric CO_2 and plant growth naturally reveals a more complex picture (in which, for example, some plant species thrive more than others and plant growth can be inhibited by other limiting factors), there are a set of very reliable processes that virtually always operate:

Increased concentrations of atmospheric CO_2 stimulate plant growth;

Plants grow by taking CO_2 out of the atmosphere;

When plants absorb CO_2 they need less water.

If plants grow better in a CO_2-rich environment, why is agriculture under threat when humans green the earth using an atmospheric fertilizer? If plants remove CO_2 from the atmosphere, why aren't climate scientists examining the power and significance of this negative feedback loop? If plants that absorb more CO_2 need less water to grow, why is this a bad thing? In short, why is CO_2 so often referred to as a pollutant?

More Disease?

At the same time that the catastrophists dismiss the likelihood that warmer temperatures will actually improve conditions for agriculture, they often worry that those same warmer temperatures will facilitate the spread of disease. Many disease vectors do thrive better in warm conditions, but so do humans. We often forget that our ability to live in, say, Chicago, is a testimony to our cultural rather than natural adaptability. Without clothes and tools and readily available fire, humans would sorely struggle to survive a Chicago winter.

Even to this day people are more likely to die from cold than from heat:

http://www.thelancet.com/journals/lancet/article/PIIS0140-6736(14)62114-0/abstract

https://www.sciencerecorder.com/news/2014/07/31/extreme-weather-kills-thousands-each-year-in-the-u-s-cdc-says/

Most life thrives in warm conditions so it is perverse to worry that global warming will make us sicker or cause domesticated crops to wither on the vine. Such a paradox certainly is possible, but to view it as probable based on little more than speculation suggests that it has been guided by the preconceived notion that warming is bad.

We emerged as a species in the winterless tropics and only later adapted to cooler environments. If we never had developed complex cultures we probably would have remained confined to our original ecological niche and surely would have failed to colonize the entire world. Is it really reasonable to think that humans would not be able to adapt to temperature increase? There is no denying that adaptation often requires both resources and resourcefulness, but both as a species and as individuals humans are constantly engaged more in adaptation than in any other activity. Much of our waking time is spent either solving or circumventing problems – is this not the essence of adaptation?

For millennia, infectious diseases were a literal and figurative plague on mankind, but human ingenuity found ways to limit their virulence and in the past century or so advanced societies have virtually eradicated the deadliness of most of them. Of course they might return if we are not vigilant, but any honest observer has to admit that they no longer influence death rates very much in advanced countries where morbidity and mortality are now governed by degenerative diseases like cancer and heart failure and stroke. We can clone sheep and genetically modify plants to our own ends, but we won't be able to cope with a few degrees of extra warmth? To make such a claim, warmists must presume that humans have very little ability to adapt to new environmental conditions. Our entire history contradicts this dark view of human potential.

Of all the infectious diseases that warmists are concerned about, the most commonly cited one is malaria. There are reasonable objections to the contention that malaria will inevitably flourish as the world becomes warmer. I will cite a number of them.

The speculation is that when the temperature increases, large areas in the mid-latitudes will become more receptive to malaria, which currently is confined mostly to the tropics. Such an idea contradicts the reality of the situation. Mosquitoes thrive not just in the tropics but even in the Arctic and sub-Arctic where they grow to such a colossal size that one might categorize them as true predators. The most serious malaria epidemic in human history occurred in Russia in the 1920's when an estimated 800,000 people were infected. The small city of Archangelsk located close to the Arctic Circle recorded 30,000 cases.

Malaria is not a purely tropical disease. It, like most forms of life, thrives better in the tropics but to suggest that warmer mid-latitude temperatures will expand its geographic range pretends that malaria cannot thrive naturally in temperate and even downright cold environments. This is simply untrue. In fact, malaria was widely spread in non-tropical parts of the world and retreated from most of those regions only when human societies found a way to eradicate it. Using the methods we used before, we could quickly and cheaply eradicate malaria everywhere – we just don't want to, it seems.

In the years shortly after World War II, DDT was used successfully in country after country to wipe out malaria and lower the overall death rate. But the wealthy countries of the West, having already eliminated malaria in their own populations, came to the conclusion that DDT was too damaging to the environment and should be banned. Thereafter, no poor countries received assistance from rich countries if they engaged in such a retrograde activity as spraying the environment with DDT. Never mind that today we use a wide array of chemicals that are much more toxic than DDT. We have decided that DDT is bad by definition and no matter how effectively it may save lives its negative effects are so, so terrible that we must never use it. A similar irrationality would be to conclude that because radioactivity is highly toxic to humans it should absolutely never be used in any medical treatments. This perspective on malaria and DDT can be read at the following conservative URL. Although the website has a political agenda, this article contains a great deal of specific, factual information that can easily be checked for accuracy:

http://www.discoverthenetworks.org/viewSubCategory.asp?id=1259

Those who worry that warmer temperatures will cause additional deaths from malaria in the future are often the same people who resist saving lives from malaria in the present using DDT.

The World Health Organization has put its weight behind the idea that warmer global temperatures will lead to higher incidences of malaria:

http://www.who.int/globalchange/environment/en/chapter6.pdf

But when you read why this is expected you discover it is based on nothing more than output from models that predict the future – models that obviously rely on unrealistic assumptions such as no concerted global effort to deal with malaria, no application of known preventatives like spraying with DDT, and no pursuit of new approaches to inhibiting the vectors of malaria.

Instead of relying on output from models, investigators should be trying to document with actual field data where temperature has increased in the recent past at the same time that the incidence of malaria has risen. This might be a good starting point for a respectable theory. Without such data, models are little more than speculation dressed in mathematical respectability.

Any theory that relies primarily on nothing more than a prediction of the future – no matter how appealing the logic of the theory or the reputation of the theorizer – is a weak form of science. Indeed, the World Health Organization tacitly admits that their model results are not very robust: their model predicts that a 2-3 degree Centigrade temperature increase would increase the number of people "at risk" of getting malaria by 3-5% and yet in their table of environmental changes deemed to affect the incidence of specific infectious diseases they do not include temperature increase as a cause of more malaria risk. They know their model is not real science so they avoid listing global warming as a cause of increased malaria.

Seen any Hurricanes Lately?

In recent years there has been a growing chorus of agitated voices singing out about how global warming is causing extreme weather events to be both more frequent and more intense. The data establishing such a proposition are never presented and even the theory that would explain the process lacks any real specificity. This is obvious because it doesn't matter what the nature of the extreme weather event might be. It could be drought or heavy rain, a heat wave or a cold snap, a hurricane or tornado or a stoppage in the locally prevailing breeze – any of these tends to get labeled as a consequence of global warming, even though we all know that each of these various conditions is caused by a totally different set of atmospheric dynamics. But do we ever see an articulated theory for why each separate one of these obviously different processes is caused by a warmer atmosphere? No, we do not. Instead we are invited to believe that global warming can cause them all and then whenever any one of them happens we are asked to accept that global warming is at work.

All educated people know that an individual weather event does not by itself offer even circumstantial evidence that a specific thing caused it (in this case, global warming), and when contradictory weather events (like drought vs. persistent rain) are proclaimed as having that same cause we become even less convinced. There needs to be a concrete, coherent theory for how global warming causes each of these different types of weather events and there have to be actual data establishing that each type of weather event has become more frequent and more intense as global warming proceeds. Neither of these scientific requirements has been met.

In fact, published studies have shown that, at least for the United States, the frequency and the intensity of hurricanes, tornadoes, heat waves, cold spells and prolonged rainfall have not increased in recent decades. Reputable scientists have concluded that so far there is no evidence that extreme weather is becoming more common. Global warming is being blamed for something that isn't happening. The following URL includes charts showing that extreme weather events have not increased in recent years:

https://wattsupwiththat.com/reference-pages/climatic-phenomena-pages/extreme-weather-page/

There are some data to suggest that extreme droughts may be linked to global warming, but even for this type of extreme weather the evidence is weak:

http://www.nytimes.com/2014/02/17/science/some-scientists-disagree-with-presidents-linking-drought-to-warming.html?_r=2

One would think that before an explanation of changes in the incidence and severity of extreme weather is even attempted, some sort of carefully controlled study would have already established the reality of the change. The fact that global warming theorists are pointing to individual examples of extreme weather events and invoking global warming as its cause is strong circumstantial evidence that they are not objective observers about what is going on in our atmosphere.

Anyone who is skeptical about the scientific integrity of this supposed linkage between warming and extreme weather naturally suspects that the more or less coincidental shift in preferred terminology from "global warming" to "climate change" is an effort to divert attention from the fact that actual global temperature has remained quite stable for the last two decades. For the "climate change" label to remain reputable its users need to do more to document the reality of more extreme weather.

Even if the incidence and intensity of extreme weather events were increasing in recent years (which was not the case in the United States), the trend would have been occurring at a time when the pace of global warming was slowing to a near standstill. Why would extreme weather events happen more frequently when the pace of global warming is slowing?

Many would argue that extreme weather is associated more with the absolute level of global temperature than with the process of global warming, but this only becomes persuasive when solid scientific studies can be cited that such weather events are indeed more frequent and more intense now than they were in the past. I am aware of no studies that do this.

Monkey Business

There is some possibility that the raw temperature data used to estimate global average temperature have sometimes been modified in ways that provide numerical support for the theory of global warming. The original raw data is not readily available to the general public whereas modified and refined data is. The modified and refined numbers are now referred to as the official numbers.

Modification of raw data is not as sinister as it sounds since the practical utility of refined data often is much greater. Here is an example: if one wished to calculate the decline in average temperature that occurs at higher latitudes, it would be useful to adjust the original raw data from each station so as to eliminate the roughly 3.5 degree decline in temperature that occurs for every 1000 feet of elevation above sea level. Conversely, to measure temperature decline with increased elevation it would be sensible to adjust the raw data so as to control for the latitudinal positions of the weather stations.

Obviously, refining raw data can be a useful way to improve its value for answering a specific question. But: refinement without reference to a particular question serves no purpose and refinement for the purpose of addressing a clearly specified question would generally make the data less useful for answering virtually any other question. This is why it is vital that authorized agencies always make readily available the original raw data. Only the potential user knows what the use is going to be, and thus it is only the potential user who can judge what sorts of adjustments to the raw data ought to be made.

An important point: most adjustments to raw temperature data do not make the data more accurate. It is very, very hard to improve the accuracy of original, recorded data; one can generally only make it more useful for answering a particular question. To tell your companion standing at the top of Mount Everest that, "No, no, it is not anywhere near this cold because if we were at sea level the thermometer would be reading about 100 degrees warmer" will not persuade him to shed some of those bulky clothes. Adjustments do not make data more accurate – only more useful for some particular purpose.

It is admirable that NOAA should refine the raw temperature data from individual weather stations so as to make it more useful for examining the most commonly posed questions about how different locations might be compared to each other through time, but the principles of science demand that the refined product not be treated as the original information collected. But this is what NOAA has done: the actual raw data no longer is readily available to the general public – only the refined figures.

The process of making adjustments to the raw data is referred to as "homogenization" and it affects many, but by no means all, of the weather stations used to compute the global average temperature. There are a number of different problems that get addressed by homogenization. I'll mention just two. First, cities tend to develop what is called an "urban heat island," a noticeably warmer temperature regime than that existing in the surrounding countryside. Second, weather stations with long and continuous temperature records sometimes get relocated or adopt new temperature recording technology, either of which can cause a sharp change in the raw temperatures being recorded before and after the event. (Note, incidentally, that weather stations in poorer countries may never have adopted the new temperature recording technology and thus may not have had their raw data similarly adjusted.)

In these two cases, homogenization proceeds by finding other nearby weather stations that did not have the same kind of distortion and then using the average trend pattern of those proximate stations to numerically modify the problematic one. As you can see, the homogenization process cannot help but be messy and complicated. It certainly is not foolproof but it is judged by most to be better than just using the raw data, and therefore necessary for the study of average global temperature.

The process of doing homogenization requires the formulation of standardized protocols so that all stations get subjected to the same set of rules regarding whether a homogenization procedure should be done and what the steps will be for doing it. Although this helps insure that judgmental bias does not play a role in the adjustments made to stations it cannot guarantee total objectivity. The rules may be "reasonable" and "agreed upon by consensus," but they nonetheless set what are to some degree arbitrary limits and boundaries and thresholds that in and of themselves cannot be judged as "correct." Whether or not a different set of chosen values for doing homogenization would yield up different conclusions regarding the average global temperature is not known – although a reasonable person would expect not.

Reasonable people often are wrong, however, and this highlights the fundamental difference between raw and refined data: for all its flaws, raw data is real whereas refined data is not. When, therefore, news reports announce that 2014 was the warmest year on record the conclusion presumes that the excellent minds and objective temperaments of those individuals who homogenized the raw data yielded up modified temperature figures that can be relied on as a better representation of reality than the very stuff they used to make their judgments in the first place. Once temperature data has been homogenized it really no longer is "data" in the honest sense of the word.

We can accept that climate scientist have good minds, but are they truly objective? According to many reports they are not. If indeed it is true that the great majority of them agree that anthropogenic global warming is real then they have made up their minds about the science and will on average tend to pay attention to any information that supports their belief even as they tend to neglect information that undermines it. This is not saying anything terrible about climate scientists; this is just human nature, as many rigorously controlled scientific studies of human behavior already have established.

Since this august body of scientists already believes in the seriousness of global warming, those from the body who homogenize temperature data will feel confirmed in their belief if the adopted procedure for homogenization yields up stronger evidence of the warming trend, and will feel somewhat dismayed if it does the opposite. These are the people, mind, who establish, and then frequently adjust, the rules of homogenization. Human decency is not the issue here: I should imagine that virtually all the involved scientists do their best to do what is right. But the circumstances are ripe for the introduction of investigator bias. By its nature, the homogenization process is judgmental and those in charge of doing the judging are invested in what the final product says. Skepticism in the face of such a situation is quite reasonable.

Since lack of objectivity is a persistent and pervasive source of error in science would it not be sensible for all the reputable temperature models to also be run using nothing but raw data? Given the urban heat island effect, it is likely that the raw data would show even more global warming than the homogenized data, and so providing the comparison might give global warming advocates a good, fact-based argument to use against the skeptics. Alternatively, if the raw data forecasts showed relatively less global warming then the custodians of the homogenization process would feel compelled to more carefully examine the issue.

At the very least, the original raw data on temperature should be as readily available to everybody as the official figures that now contain a host of modifications.

Sporadic controversies have arisen in which individuals outside the mainstream climate science community have claimed that the raw temperature data for individual weather stations have been modified in ways that systematically shift their long term trends so as to favor the global warming agenda. In each instance there have been responses from the establishment that purport to show that the challenge is specious – usually by presenting data that suggest the adjustments result in as many trend shifts in the cooler direction as in the warmer one. The following article, which recently appeared in *The Australian*, discusses the growing list of controversies surrounding the homogenization of temperature data:

http://www.thegwpf.com/inquiry-into-australian-temperature-data-adjustments-begins/

Many in the global warming camp view these controversies as the misguided imaginings of individuals with an axe to grind. The problem with this narrative is that the global warming skeptics are a motley collection of marginalized academics who have rarely received any compensation for their troubles, have certainly not become famous for resisting the mainstream, and have frequently been blackballed or silenced on account of their contrarian views. I do not believe that a single skeptic has yet been shown to have gotten rich or famous by expressing skepticism, although there are a number of global warming believers who have profited handsomely from their advocacy (Al Gore is the prime example).

Controversies about homogenized data are cropping up based on temperature records in lots of places. Each new controversy seems to focus on a different part of the world but the issue always stays the same: Why are homogenized records showing more significant warming than those based on raw data?

In the United States, a climate blogger named Steven Goddard (a pseudonym) contends that the raw temperature data for the United States indicates that the hottest years on record occurred in the 1930's and not in the 21st century. I will not pass judgment on the efficacy of Goddard's claim, but imagine how easy it would be to settle this matter if the original raw data were readily available. Here's the URL for Goddard's post:

https://stevengoddard.wordpress.com/2014/06/23/noaanasa-dramatically-altered-us-temperatures-after-the-year-2000/

A different blogger named Paul Homewood claims to have accessed both the raw and the homogenized data for all relevant weather stations in Paraguay. His comparisons indicate that for all nine stations, including the rural ones, the homogenization process has shifted the temperature trends towards more warming:

https://manicbeancounter.com/2015/02/08/is-there-a-homogenisation-bias-in-paraguays-temperature-data/

Once again, the issue is easily evaluated if others are able to compare the raw and homogenized data.

A similar charge has been made regarding Arctic weather stations, following which a mainstream scientist dismissed the amount of additional warming as trivial, only to be rebutted by a different blogger who using the same data calculated it to be ten times as great as the mainstream scientist claimed:

http://www.breitbart.com/london/2015/02/07/breathtaking-adjustments-to-arctic-temperature-record-is-there-any-global-warming-we-can-trust/

In Australia, the same story has emerged, largely as a consequence of data snooping by a blogging biologist named Jenifer Marohasy. She claims to have found a host of Australian weather stations for which the homogenized temperature data show substantial warming when the raw data identifies cooling, no trend, or only slight warming:

http://www.breitbart.com/london/2015/02/07/breathtaking-adjustments-to-arctic-temperature-record-is-there-any-global-warming-we-can-trust/

Figure 11 is taken from this URL and shows the kind of change that has been wrought via homogenization for Amberley in Queensland.

These four controversies regarding homogenization are not the only ones; there are more. If you look around online you can easily find additional information about any one of them, including rebuttals from the mainstream science community. The intent here is not to insist that the climate science establishment is cooking the books (although it may be). It is to point out that this whole business of doing homogenization is now under constant assault. Convenient access to all the raw data in a readily usable form would create the transparency that has always been viewed as essential to good science. When such information is tucked away in obscure locations with complex protocols for downloading and using it, it is only natural for some people to be skeptical about the real intent of homogenization procedures.

Whether or not homogenization is being used to give the appearance of more atmospheric warming than the raw data would indicate, there is always the risk that homogenization procedures will be flawed in some way that biases the temperature record. A couple years ago, a peer reviewed article in a major scientific journal established that precisely this has been happening. The article showed that the statistical procedure for adjusting temperature data when a city weather station is moved to the suburbs "can lead to a significant overestimate of rising trends of surface air temperature." This highlights the observation made earlier that no matter how well conceived a homogenization procedure, it may contain flaws. The fact that, once again, the flaw results in a more pronounced pattern of warming only confirms to the skeptic that there is good reason to be skeptical about the entire homogenization procedure (an abstract for the article and a commentary regarding the article written in more accessible English can be found at the following two URL's):

http://link.springer.com/article/10.1007%2Fs00704-013-0894-0

http://variable-variability.blogspot.com/2014/02/effect-of-data-homogenization-on.html

By now it should be evident that although homogenization may be appropriate in theory it also may lead to results that are either intentionally or unintentionally biased. With this as background, let us momentarily consider the implications of Climategate, the series of private e-mails stolen from the Climate Research Institute at the University of East Anglia back in 2009:

https://en.wikipedia.org/wiki/Climatic_Research_Unit_email_controversy

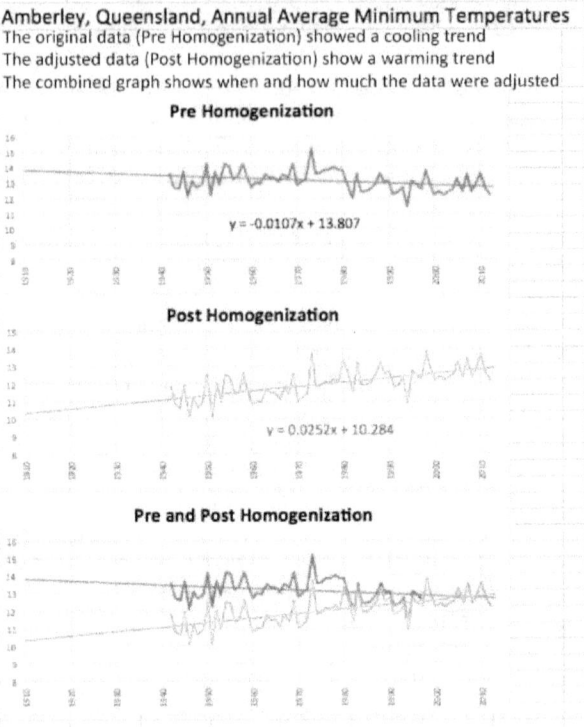

Figure 11

Amberley, Queensland, Annual Average Minimum Temperatures
The original data (Pre Homogenization) showed a cooling trend
The adjusted data (Post Homogenization) show a warming trend
The combined graph shows when and how much the data were adjusted

When the e-mails were made public, they exposed a number of very prominent climate scientists as being strongly committed to the human-caused global warming idea. They wrote e-mails that suggested they might have no problem with obscuring the fact that global temperature has stopped rising, with working to insure that climate skeptics not be published in certain journals, and with presenting data in such a way as to lend credence to the global warming agenda. Various government committees absolved them of any wrongdoing, but anybody who reads the e-mails immediately realizes that these individuals are uncomfortable with challenges to their ideas. All of us are, but scientists have a solemn obligation to be open to contrary points of view. If they are not, they cannot function as true scientists.

Many global warming advocates dismissed Climategate as a trumped up scandal, but the attitude projected by those scientists in their e-mails showed a much greater concern with protecting a particular point of view than with getting at the truth. This is just human nature but any student of science knows that it has the potential to obstruct or derail the progress of science for at least a short time. Even though the scientists caught up in Climategate were absolved of any serious wrongdoing, the tone and content of their communications certainly justifies an objective outside observer having some reservations about their intentions. Skepticism is a perfectly reasonable reaction to any scientific conclusions advocated by scientists who use questionable methods to promote their beliefs.

The Precautionary Principle

Because a yawning gap has emerged between temperature predictions and reality, global warming apologists have been motivated to invoke the precautionary principle as an appropriate course of action for humanity. This principle holds that even when an adverse future condition is less than certain, it is prudent and sensible for humans to plan as if it will occur. The thinking is that by taking preventive action in advance it may be possible to ameliorate or avoid the consequences of the event at a moderate cost – and if the event fails to materialize the preventive actions will have been a small price to pay.

The IPCC advocates implementation of the precautionary principle as a way to counter the catastrophic consequences of global warming precisely because neither the warming itself nor those extremely negative consequences are a certainty:

http://www.ipcc.ch/ipccreports/tar/wg3/index.php?idp=437

The IPCC admits that it is not even sure a high level of global warming will materialize but it nonetheless advocates for immediate precautionary action to combat the adverse effects of global warming. In short, the IPCC wants to implement the precautionary principle by forcing the world economy to abandon its reliance on fossil fuels – not in a century or two but within the next few decades. Any economically literate person understands that this would be a revolutionary change in an absurdly short period of time. And it would be naïve to believe that such a revolution might be achieved without coercion. This IPCC is making a blatant play for political power, justified on the basis of a crisis that – because it cannot really be proved to exist – has to invoke the precautionary principle.

As all this makes clear, invocation of the precautionary principle implies that its advocate is not so certain that a specific future scenario will indeed play out. As soon as I hear mention of the precautionary principle I ask myself why is a discussion of this sort necessary if in fact the science is settled and a simple explanation of the facts of the matter could prove it.

The precautionary principle is an admission that the science is really not so settled. If the science were certain then there would be no need to resort to such a weak line of thinking. The most obvious parallel to the precautionary principle is Pascal's wager: 'If you believe in God and there is no God then no harm done, but if you don't believe in God and God exists then you will pay a big price; thus it is better to believe than to not believe.' The issue of belief is the critical factor here. Are we to judge the merits of the global warming theory based on whether belief in it will yield practical benefits? Would it not be more rigorous and more scientific to judge the merits of the global warming theory by subjecting it to objective analysis and evaluating the degree to which the theory matches reality?

I would argue that the precautionary principle is inappropriately applied to the global warming issue because we do not clearly understand the context within which it is applied. We have no seriously derived estimate of the probability that catastrophic global warming will occur and we have inadequate estimates of the costs associated with preventive vs. remedial approaches to dealing with the problem.

The IPCC claims in its most recent report that catastrophic global warming is more than 95% likely, but this is obviously an effort to put a scientific façade on something that nobody has a reasonable way of measuring. The IPCC advances this measurement of certainty with no indication of how it was arrived at. In other words, this supposed statement of fact cannot be verified by an independent observer using the same data and methodology, and thus is not based in science.

Without some verifiable method for estimating the certainty of an event, it is virtually impossible to determine when the precautionary principle should be activated and when it should not. Consider the theoretical example of a large asteroid hitting earth. There is a fair amount of agreement that such an event would be catastrophically disruptive of global life. There is also virtually unanimous support for the notion that it will happen sometime in the future. The problem is that we don't know when the event might occur and we only have a purely speculative sense of the nature and extent of the ensuing catastrophe. We have, furthermore, a very poorly developed methodology for evaluating the costs associated with either a preventive or a remedial approach to the problem. And yet, if application of the precautionary principle were to be taken seriously we would need to start doing something now.

For those who view global warming as a serious problem, the predictions put out by the International Panel on Climate Change are viewed as near certainty, and consequently the only significant remaining question is, "What are we going to do about it?" Such people have already decided that the global warming crisis is real and eminent and serious. They need to go back and research exactly how the IPCC arrived at the conclusion that catastrophic global warming is 95% certain – and when they do they will discover that this quantification of probability has no scientific basis.

Espousal of the precautionary principle is disingenuous because those who advocate for it would not use it to judge the appropriate course of action for responding to some event that they feel is unlikely (like the errant asteroid). They use the argument as a tool of persuasion: "I may be wrong, but logic dictates that that you should take my advice anyway."

This approach to the understanding of truth is by its very nature unscientific. Why is it necessary to invoke the precautionary principle if the scientific case for global warming is so iron-clad? The most likely answer is that the advocates of precautionary action, so convinced that they are right about the eminent threat, view it as the only way to win over to their own point of view those who are too ignorant to understand the science. Does the precautionary principle crowd really think that skeptics are too insensitive to recognize the latent paternalism in such an approach?

Application of the precautionary principle to the possible consequences of global warming is particularly ironic considering the widely accepted and yet generally ignored belief that the world may be at the high point of an interglacial period. We have solid evidence that over the past 400,000 years there have been four major episodes of glaciation during which global temperatures were much colder than today. Each of the four lasted for roughly 80,000 years and each of the first three was followed by a comparatively brief episode of ice cap retreat before the onset of a new advance, an interglacial period of 10,000-20,000 years duration. The fourth glacial episode only gave way to an interglacial starting about 12,000 years ago and the temperature regime since then seems to be comparable to what happened during the warming phase of the three preceding interglacials.

The relatively warm global temperatures of today are similar to the peak warm periods of the three preceding interglacials so there is an historical pattern supporting the notion that we may soon see global cooling that leads to another glacial episode. We even think we understand why the world has been experiencing these glacial epochs with their brief interglacials and so the recurring pattern surely argues for implementation of the precautionary principle.

Should we not do something now while we still have time? After all, another glacial advance that lasts for 80,000 years would be at least as disastrous for life on earth as the amount of atmospheric warming that global warming advocates predict. Imagine, for example, all of Canada and the northeastern part of the United States under ice. Imagine the Rockies and the Sierras submerged in ice. Imagine the remainder of the 48 states as either Arctic tundra or northern boreal forest. Surely this scenario is as unsettling to our way of life as a sharp rise in atmospheric temperature with its attendant invasion of subtropical and tropical climate regimes and its associated sea level rise.

Figure 12
Climate Record past 450 Thousand Years

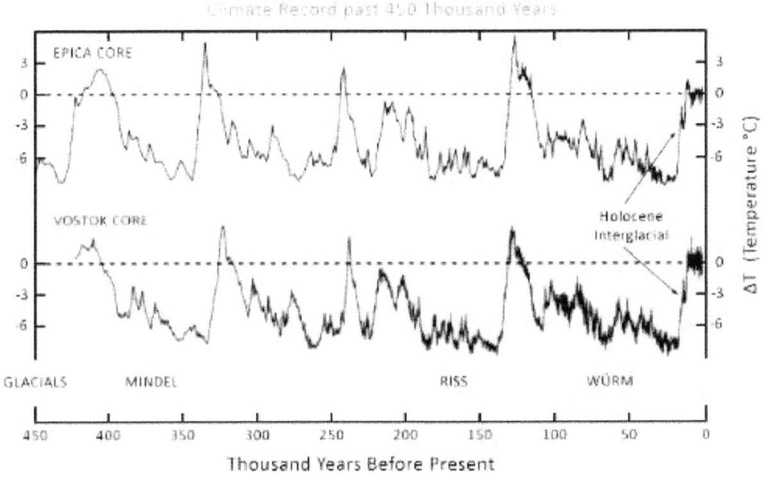

Using ice cores from ice sheets, it is possible to reconstruct estimates of temperature change during preceding millenia. The two reconstructions shown here are so similar that it lends confidence that they may be accurate.
The two graphs make it quite clear that the temperature at present is typical of an interglacial period and that if so then the near future may bring colder conditions.

Sea level disaster, incidentally, works both ways. If we were to enter a new ice age comparable to the last one we should expect a sea level drop of around 400 feet, and early on in the process virtually all port cities in the world would be obliged to abandon their main reason for existence. All existing sea level canals would become unusable, including Panama and Suez. Virtually all rivers would be subjected to great changes that would require substantial adaptation by any settlements situated next to them. In short, the implications for human life on the planet are no greater for global warming than they are for global cooling. There is little to choose between the two when it comes to application of the precautionary principle.

Because those who advocate the precautionary principle wish to employ it even if the potential disaster is unlikely (a conclusion forced by the fact that they encourage skeptics to adopt it even when those skeptics don't think the disaster will happen) then consistency obliges them to advocate it for dealing with the perfectly reasonable proposition that we might be about to enter a new ice age. This means they would have to employ it against the prospects of both global warming and a returning ice age. They will be hard put to avoid an enormous expenditure of resources to achieve contradictory ends.

It is true that right now global temperatures are going up, just as they did do for about 10,000 years following each of the other glacial periods, but so far we have seen no sign of the acceleration in global warming that the climate models project. In case the global warming scenario is wrong and the ice age cooling scenario turns out to be right one we might do better to wait a little longer before committing our reserves to the battle.

THE SCIENTIFIC CONSENSUS

"Settled Science" Does Not Exist

The phrase "settled science" contradicts itself. Science proceeds by rejecting untruths that experiments reveal for what they are: falsehoods. Science does not have the capacity to prove something is true – it can only confirm that reasonable alternatives are false. In other words, science makes a highly sophisticated estimate of what the truth may be by eliminating alternative possibilities, but cannot ever achieve absolute certainty that a remaining possibility is in fact the answer. Given this inherent limitation on the scientific method, the best that can be achieved is a diminished degree of "unsettledness."

For this reason, skepticism always has been regarded as essential to the scientific method. Nothing is ever known for sure and thus must – often and forever – be subjected to challenges. Whenever a claim is made that a particular scientific principle no longer is in doubt, well, that is a sure sign that the spokesperson is abandoning the realm of science.

Of course in the practical world decisions must be made that presume a particular scientific principle is indeed the truth, but to simply denounce those who challenge the prevailing paradigm is to launch an *ad hominem* attack that is unworthy of true scientists. The obvious and appropriate course of action is to counter the skeptic's arguments with logic and analysis that refute the relevance of data inconsistent with the theory.

If the skeptic has also introduced evidence to suggest a different explanation for the phenomenon then efforts must also be made to show why the accepted understanding is better supported by available facts than the skeptic's understanding is. In any event, those who dismiss the challenge based on nothing more than the flawed character of the skeptic are abandoning science and instead attempting to justify their own beliefs (and beliefs are never science).

If the case for human-caused, catastrophic global warming is solid then those who advocate for it should have no problem countering any data-based challenges brought by skeptics. But usually, the mainstream reaction to a skeptic of the global warming paradigm is to disregard the skeptic's arguments and instead claim that that person is guilty of some undesirable human failing such as greed or attention-seeking or some form of insanity. But this will not do: the scientific method demands that the discussion be about the facts of the case rather than the reputability of the individuals who are presenting the facts.

The facts are that various aspects of the global warming theory are challenged by hard data. There are of course data that circumstantially support the theory but that is not the point; the point is, there are inconsistencies in the theory that must be explained if the theory is to survive. All the supportive data really only show that some alternative ideas may not be true. That data does not prove the theory is correct. Any significant body of contrary evidence can destroy the theory if it is not adequately explained.

Here are some of the major questions that skeptics think are being left unanswered – questions that arise because there are data suggesting an inconsistency in the theory:

> Why have all the reputable models forecast more rapid global temperature increase than has actually happened?

Why do model forecasts predict more rapid warming in the upper atmosphere than near the earth's surface when contemporary data indicates that that is not happening?

Why do the models predict a more intense heat build-up in the upper atmosphere of the tropics when accepted data establishes that there is no such heat build-up?

Why do the models predict a proportionate decline in heat loss to outer space as the atmosphere gets warmer when at least one reputable study indicates the opposite?

Why do the models forecast an unstable acceleration of atmospheric heating considering that no such event occurred during geological time periods when CO_2 levels were much greater than today?

Since sea level rise has been very small in recent decades (about three millimeters per year), how can net wasting of the Greenland ice cap possibly be significant in scale?

Since the amount of global sea ice has remained reasonably constant for decades, how can global warming of the atmosphere be responsible for melt back of sea ice only in the Arctic?

Since data indicate that during the time of the Vikings global temperatures were higher than today, and yet CO_2 levels were much lower, how is it possible to contend that atmospheric CO_2 is the only reasonable explanation for global warming?

Since for hundreds of thousands of years CO_2 and temperature have had coinciding patterns of rise and fall, but with temperature change occurring first, how is it possible to contend that in the natural environment (and not the laboratory) CO_2 manipulates temperature instead of the other way around?

Since over 95% of all CO_2 emissions into the atmosphere are a result of natural forces beyond our control (and even our understanding), why is the small human contribution referred to as pollution?

Since global temperature has already gone up and down a lot during the few millions of years of human existence, why is the possibility of a sharp temperature rise now considered to be an existential threat?

Since no reputable studies have yet documented the reality of a link between warmer temperatures and extreme weather events, why are we encouraged to believe that such things as hurricanes and heat waves are the consequence of global warming?

Since "global warming" refers to a very specific physical process believed (by alarmists) to be the driving force behind such things as sea level rise and extreme weather events, why should we forsake the label "global warming" for the vague and meaningless label of "climate change"?

Since virtually all climate scientists get paid by governmental entities to do the kind of research those governments want done, why are those scientists considered to be objective and disinterested researchers?

Since science relies on facts rather than opinions, what does it mean to call it "settled"?

Many, many more such questions can be posed that challenge the global warming narrative, but this list at least gives a sense of why skeptics think that the global warming theory needs to be more thoroughly vetted before being regarded as "settled science."

The skeptic need not prove that some alternative explanation is correct. If data contradict what the theory would predict then the theory must either adapt or die. Even the lack of a good alternative explanation would not be sufficient to keep the theory alive. That is the reality of science.

A belief in global warming may rest on confidence in a set of scientific principles, but climate scientists with an agenda are violating those principles. A temperature reading may be a fact but science is the process of deriving conclusions from facts – and that process yields promising results only when certain rules are followed, in particular:

1) Data trumps theory.
2) Absolute truth is unattainable.
3) Commitment to an idea inhibits the pursuit of scientific truth.

There is of course much more to science than this but these surely are among the guiding principles of science and in all three cases the contemporary global warming advocate is in violation.

"97% of All Climate Scientists Agree . . ."

Even though scientific consensus does not contribute a whit to the certainty that a specific scientific idea is true, it usually does give politicians and policy makers their most reasonable basis for judging the matter. Decisions always get made based on incomplete and imperfect information, so on the surface the contention that the global warming theory needs to be viewed as "settled science" is perfectly reasonable. But in the global warming debate the contention is that the matter has been settled because 97% of all climate scientists agree that humans are causing catastrophic increases in the accumulation of greenhouse gases (especially CO_2). The problem is that there is no such agreement.

So where did the idea come from that 97% of all climate scientists are in agreement? It comes from a single article written by John Cook and others, and published in *Environmental Research Letters*:

http://iopscience.iop.org/article/10.1088/1748-9326/8/2/024024/pdf

The article was entitled "Quantifying the Consensus on Anthropogenic Global Warming in the Scientific Literature." It concluded that 97% of all climate scientists agree that anthropogenic global warming is happening, but its methodology was not sound.

The authors of the article identified nearly 12,000 academic articles, published between 1991 and 2001, that were located using key word searches under "global climate change" and "global warming." Each article had an abstract of its contents and their study looked at whether these abstracts expressed any opinion about the idea of anthropogenic global warming. Almost exactly two thirds of the article abstracts expressed no opinion whatsoever, while the remaining third (roughly 4,000 articles) did contain some sort of opinion statement. Within this remaining third of all studies, about 97% of the abstracts expressed some sort of support for the idea of human-caused global warming and only about 3% expressed some other view.

The authors of the article categorized statements made in the abstracts according to the degree to which they supported the idea that humans are causing global warming. A mild level of support would be a statement indicating that humans are contributing (but perhaps only insignificantly) to warming of the globe (which may be minor); a strong level of support would be a statement indicating that humans are the main cause of global warming, which in turn is significant in its degree and consequence. The vast majority of the abstracts only expressed very mild support for the AGW (anthropogenic global warming) theory.

No matter how faintly or reservedly an abstract expressed agreement with the notion that humans are contributing to global warming, it ended up categorized as supportive of the proposition that humans are the main cause of atmospheric warming and that this is going to be a serious problem for the world.

But of course no educated person is going to deny that humans put some CO_2 into the atmosphere and that CO_2 contributes to atmospheric warming. This is not the issue. The issue is: "How much? And how serious is it?" Virtually all skeptics are skeptical about the significance of the problem, not about the fact that CO_2 emissions into the atmosphere cause some warming. Any skeptic who wrote one of the article abstracts could very easily have expressed a view that would have ended up being categorized as supportive of the human-caused global warming narrative.

Coopting support for an idea in this manner is fundamentally dishonest. It would be like advocates of "pro-choice" in the abortion wars claiming as a supporter anybody who accepts that in the case of rape or threat to the mother's life an abortion might be permitted. Using such an approach, the pro-choice camp could claim that the great majority of the US population supports their position when in fact that is not the case.

The Cook article also uses a suspect methodology when it comes to the question of non-responses. Remember that two thirds of the reviewed abstracts expressed no view regarding the idea of human-caused global warming. To simply discard these abstracts from the analysis was tantamount to presuming that their composite view of the issue was comparable to that found in the abstracts expressing a view. But if authors of articles having to do with climate change fail to express an opinion about the topic would it not be reasonable to think that they are skeptical about its significance? After all, their articles pertained to the issue of anthropogenic global warming and yet in their abstracts they refrained from making any judgment about it. This suggests to me that they are not so taken with the idea and, even more importantly, do not think their research supports the idea.

The IPCC Juggernaut

Although that one published paper became the basis of the claim that 97% of all climate scientists agree about the threat from human-caused global warming, the reputability of the scientific consensus argument also relies on the supposed objectivity of the International Panel on Climate Change (the IPCC).

Established in 1988 by the United Nations, the IPCC is charged with developing reports that assemble and interpret factual information about human-induced climate change, how it will affect the world, and how the world might respond to the anticipated climate changes. The IPCC enjoys a great deal of public trust, partly because in 2007 it (in conjunction with Al Gore) received the Nobel Peace Prize for its efforts to educate the world about the climate change issue.

The IPCC only considers <u>human-induced</u> climate change; its concern with natural causes of climate change is limited to those facts that might in some way relate to or interact with the human-induced processes. In other words, the IPCC takes as given that humans are significantly altering the global climate and, by default, tends to discount any naturally occurring dynamics. Most global warming skeptics have the opposite perspective: they take as given that natural forces are the primary causes of climate change and that a significant human influence on climate has not yet been proven. Global warming skeptics question the value of the IPCC work because it presumes a level of human influence that has no real basis in data. To this day there still is no reputable methodology for scientifically establishing the proportionate extent to which global warming is caused by humans, and so, from the skeptic's point of view, the entire IPCC edifice is built on a shaky foundation.

One issue and only one issue drives the IPPC agenda: the degree to which human activities are contributing to the accumulation of CO_2 in the atmosphere. The entire IPPC case for human-caused global warming relies on the notion that our burning of hydrocarbons is a critically important addition to the atmospheric CO_2 – so important that it will soon take us past an arbitrarily designated "tipping point" beyond which global warming will spin out of control.

Given that global temperature is known to have fluctuated to both much higher and much lower levels in earlier times (before significant human contributions to atmospheric CO_2), is it not reasonable to question the unproven tipping point to which the IPCC and many global warming advocates often refer? Without a very specific theory and a body of supporting data, the idea of a tipping point is suspiciously like the common advertising ploy that "supplies are limited and if you do not order right now you probably will lose out." We all readily see through this ruse when advertisers use it, but when a Nobel- prize-winning entity makes a similar sort of claim we hesitate to challenge the idea that urgent action is necessary.

The IPCC is handicapped not just by its mandate to only consider human-induced forms of climate change; it is also crippled by its very structure. Its data and information about the nature and degree of human-induced global warming is compiled and collated by a very large number of the most reputable scientists in the world, but the analysis and interpretation of that data – although initially done by those same respected scientists – is edited and codified into a final summary by a group of non-scientists who represent governmental bodies and have no professional obligation to arrive at objective conclusions. The result is that each voluminous IPCC report contains a great deal of reliable information about what science has say on the subject but then attaches summary conclusions that are, to put it bluntly, shaped by political interests. The result is executive conclusions that often – indeed, usually – are not supported by the facts given in the rest of the report. Unfortunately, reporters and interested readers rarely look beyond the executive summaries when trying to ascertain what the IPCC report has to say.

This structural arrangement means that large numbers of the world's greatest scientists contribute to the assembling of an IPCC report, only to see their work "interpreted" by the political hacks who draw conclusions from it. Not only that, the IPCC is then able to claim that its summary conclusions are supported by the world's best scientists, even though those same scientists had a limited – one might say merely advisory – role in drawing conclusions from their work. A number of these scientists are perfectly happy with the conclusions that get drawn from their work, but a fair number are much less so. Nonetheless, the IPCC is able to imply that all contributing scientists, including the disgruntled ones, are supportive of the human-induced global warming idea. Once again, consensus pseudoscience gets substituted for real science.

Pseudoscience in a Tux

A large number of the world's most highly respected scientific institutions have expressed support for the idea that human-caused global warming is a real and urgent threat to the world community. The list of such supportive bodies is extraordinarily long and cannot be dismissed as trivial. No fair evaluation would accord to them purely base motives for the position that they have adopted. Their reputations precede them and their dedication to the principles of science do not deserve to be belittled. For those of us who regard their expertise as much superior to that of the ordinary non-scientist the weight of their opinion cannot be shrugged off lightly.

But at the same time, their baldly stated attitude of belief is a clear expression of a non-scientific idea. Belief has no place in science. To believe in something is to violate the principles of science. Belief is by definition an acceptance of a proposition regardless of whether facts support it. That scientific bodies as institutions formally express belief in the climate change idea is evidence that they do indeed think the data support the belief, but the act of believing requires an abandonment of the scientific method. The only acceptable attitude in science is that of skepticism: belief is taboo.

That respected scientific institutions should endorse a particular belief is an indication that ideas about the way the world ought to be have begun to interfere with the uncompromising pursuit of an understanding about the way the world is. Science is a harsh taskmaster and will not tolerate such emotionalism. It is fair enough to labor for a particular cause, but as soon as one does so there can no longer be any pretension that the labor is an expression of the scientific pursuit of truth.

People believe things for a great variety of reasons but once a belief is adopted the only true test of its durability is the capacity to continue believing in spite of evidence to the contrary. Belief virtually demands that the believer dismiss contradictory evidence, and this violates the investigatory principles of science. To believe that something is true prejudices the believer against anything that challenges that "truth," and this encourages a flawed approach to scientific investigation.

Openness to contradictory ideas is essential to good science. By committing themselves to a belief in the reality of human-caused global warming these reputable scientific institutions have cut themselves off from the constant search for new knowledge that gradually enlightens us. They have, in effect, claimed an absolute understanding that brooks no dissent – a demoralizing abandonment of the principles of science. Just as a monk or a nun cannot engage the secular world in the ordinary way and remain true to the calling, a scientist cannot engage in activism without compromising the principles of science.

In spite of their sterling reputations, these myriad institutions have compromised themselves: their goal is to find truth using science but their formally expressed belief in human-caused global warming is in effect a pronouncement that "the ultimate truth about this issue has been found." Why would anyone continue to search for truth and understanding if certain that they already had been found? These institutions have, in effect, forsaken science for politics – something that would be no less the case if their codification of truth were that humans do not contribute to global warming.

The Dark Side

Rarely are we as humans motivated by purely altruistic or purely selfish concerns. In virtually all situations our behavior blends the two. This is not to say that some people are not more altruistic than others or that narcissism does not exist: the two ingredients in the blend are mixed in very different proportions by different individuals, and much of our judgment about good versus evil relies on how successfully self-interest has been curbed and altruism cultivated. Neither the global warming advocates nor the skeptics are immune to this law of human nature.

Just as skeptics have their hidden selfish reasons for challenging the global warming narrative, advocates have their selfish reasons for supporting it. But of course both groups profess altruistic reasons for their positions, and these should not be dismissed out of hand.

In recent years, many global warming advocates have discredited – or attempted to discredit – the arguments of skeptics as being motivated by little more than self-interest. And yet, should it not be the other way around? Should it not be the global warming advocates who get accused of self-serving behavior? They, after all, represent the mainstream. They are the establishment. They are the ones who hold the power and the influence. Are we to utterly disregard the fundamental principle that power corrupts people? Are we so ignorant as to believe that consensus cannot be abused?

Consider the situation. Who pays for research into the question of climate change? Governments do – not just the United States government but central governments around the world. Most scientific research is funded by government entities that receive grant proposals and make awards to the proposals they like. It is true that corporations and individuals and non-profit organizations in the private sector also proffer grants for scientific research but theirs is a much smaller proportion of the entire science funding pie. Governments provide the lion's share of the money.

Nobody would deny that governments have a tendency to favor research that might uncover knowledge that leads to practical applications. This is not to say that governments avoid funding pure science for science's sake; but when a government wants to get to the moon or solve the problem of poverty or reduce the national reliance on foreign sources of energy, the money available for science tends to flow towards those research proposals that offer the most promise of advancing such interests.

Scientists and universities are competing with each other for research money so naturally they look for ways to propose research likely to get funded. If they can somehow associate their grant proposal with the major governmental issue of the day they know their prospects greatly improve. Their actual science project may be – probably is – substantially motivated by altruistic pursuit of truth but their strategy for getting financial support often is not. They need the grant so they do what any sensible person would do: they structure their research proposal in such a way as to offer the prospect of useful knowledge for advancing the governmental agenda. If you are a scientist and you wish to thrive in your occupation your prospects for doing so virtually require that you play this game. Failure to do so puts you in a position very comparable to the professional baseball player or professional cyclist of years past (we hope) who refused to take performance enhancing drugs.

Most governments nowadays are committed to the idea that human sources of atmospheric CO_2 are causing catastrophic global warming and they want to do something about it. These governments have accepted that the science of global warming is settled, but their acceptance often is motivated by self- interest.

Poor countries of the world think there is the prospect that large amounts of money can be extracted from the rich countries since they, the poor countries, stand to suffer the most from possible global warming consequences whereas the rich countries have contributed much the greatest share to this CO_2 problem. For the rich countries, the CO_2 threat, like war, allows politicians to exercise a level of control over society that would be otherwise unjustifiable. In the case of international institutions like the UN, the global nature of the CO_2 issue seems like a perfect justification for moving towards world government. All these governments are in fact collections of individuals whose personal prospects rise and fall with the fortunes of their employers, and so many of these employees stand to benefit if the global warming agenda is advanced.

All this in no way proves that human-caused global warming is false, but if the ideas of skeptics are to be challenged based on nothing more than their pursuit of self-interest then it is only fair to point out that the advocates of global warming should not be throwing stones. In the end science will decide, but at present the discussion is largely confined to imputed motives and consensus politics. Only when the discussion shifts back over to a consideration of the data and the evidence for and against human-caused global warming will we will be back on a path towards truth.

HOW DID WE GET HERE?

Significant global warming is not happening. This has been true for nearly twenty years. It naturally follows that all the adverse side effects presumed to accompany global warming either do not exist or are logically the consequence of some other phenomenon. Whether rapid global warming will return in the near future is an open question. If it does it will be hard to know whether increasing levels of CO_2 in the atmosphere are the primary cause. To go one step further and confirm that human activity will be the primary source of the CO_2 increases will be even harder to determine.

The theory of global warming has not been proven wrong, but the data so far have provided scant support for it. The atmospheric temperature should be warming quickly but it is not. Sea level should be rising rapidly but it is not. Sea ice should be diminishing but it is not. Land based ice caps should be melting rapidly but they are not. Weather extremes should be intensifying but they are not. The data do not support the theory in any meaningful way and any honest scientist would have to admit that the theory must either adapt or die.

Many former believers in global warming – they might be referred to as "fair-weather believers" – have fallen away and no longer adhere to the orthodoxy. They may not have reached the point of outright rejection of the theory but they have certainly become closet skeptics. A goodly proportion of them are ordinary people with no special expertise and no special investment in the question. They know they are not in a position to challenge authority – be it political or scientific – and they understand that the sensible course of action is to abandon belief but not really tell anybody. They look at global warming and they see an emperor with no clothes on, but they are not inclined to admit their heretical point of view to their next door neighbor.

They are in the closet for the same reason that homosexuals were for so many years in the closet: societal pressure to conform is intense and unforgiving. Although assuredly less hateful than that sexual form of intolerance was, the current abhorrence of skepticism about global warming is similar in kind: it simply is not acceptable nowadays for an educated person to be opposed to the idea that human-caused global warming is a fact of life. And who wants to be considered anything other than an educated person?

To challenge the global warming theory one must be prepared to undergo reputational assaults. Many of the highly educated have launched such attacks and I suspect that many who do not launch them at least sympathize with the endeavor. Why these otherwise reputable people should feel free to behave in such an uncivilized way is an open question, but there is no reason to think that smart people are less susceptible to groupthink than anybody else.

Those who speak openly about their global warming skepticism know that they will be exposing themselves to verbal attacks and yet they do it anyway. Is this because they are going to become famous for their contrariness? Because they are being paid by big oil? Because they are anti-science? Because they had a dysfunctional childhood? These sorts of silly explanations are unconvincing. Why do facts not matter when we look at this sort of question?

Why is it that an adjunct professor of modest means named Willie Soon – a world expert on solar radiation – is labeled as an opportunist supported by big oil when he challenges the global warming theory even as Al Gore, (a non-scientist) who has become a near billionaire by riding the global warming pony, continues to be viewed as science's prophetic equivalent of Christ?

How have we arrived at a situation in which a scientific theory so lacking in certainty has become accepted by so many in society, but especially by so many who are actually involved in the scientific enterprise? Perhaps future historians will offer a convincing explanation for how this happened, but since we live in the present instead of the future I feel compelled to make a few immediate suggestions.

First, there is the unhealthy respect for science that has come to infect modern society. Science has indeed been responsible for much that is good (and bad) about the modern world, but unquestioning acceptance of what science is doing is a form of totemic worship. Science is no more immune to corruption than any other human institution and when it comes to the science of global warming there are clear signs that corruption has set in.

As with our politicians, respect should be earned, not granted automatically. This was one of those "special" attributes of our original country – a respect not for authority but for proven accomplishment. People may differ in their view regarding what constitutes accomplishment, but when scientists are identified *en masse* as the only appropriate source of opinion on a matter then respect for accomplishment has been abandoned and respect for status has taken its place. The hundreds of individual climate scientists involved in global warming studies and the thousands of studies that they have published most likely have a great many meritorious traits, but discovery of some profound new scientific truth is not among them. The fact that CO_2 in the atmosphere can cause a mild greenhouse effect was discovered by a French scientist over a hundred years ago. What has the contemporary crop of scientists added to that?

But let it also be said that even the giants in science – the Einsteins and Darwins and Newtons and Galileos – do better when their ideas are met with skepticism. It keeps them on their toes and obliges them to perform at a higher level. In this respect, it is fair to say their soaring accomplishments were achieved not just by sitting on the shoulders of the great scientists who preceded them but also by the skepticism of the great mass of ordinary scientists who felt a natural aversion to simply accepting outlandishly new ideas. It may be dissonant music to the progressive's ears, but true science will and should remain a bastion of conservatism.

Human-caused global warming is an outlandishly new idea. Most everybody experiences a thrill of excitement when first exposed to a new idea. The problem with the anthropogenic global warming idea is that – like most new ideas – it falls apart when questioned vigorously.

The second reason that I think the global warming paradigm has been unadvisedly embraced is that science has become a big business. To be a small business is not bad: humans as individuals have always been motivated to do things by the prospect of material advancement, but when humans do so in the context of a big business they are usually obliged to buy into a system that features a highly complex bureaucracy that favors the sycophant and the well-connected over the creative individualist.

Small business, on the other hand, dictates that success for the individual will only come to those who satisfy the wants and needs of the larger world. Big business also admirably fills these wants and needs but those who labor within it are not so directly exposed to its demands. As a bureaucratic enterprise, science today requires respect for those in charge, obedience to their whims, and undue commitment to cooperative effort. It is rare nowadays for a single person to publish a scientific article, and the explanation for this remarkable shift in scientific productivity is often justified as being required by the complexity and interdisciplinary nature of modern science. I will not attempt to directly challenge this standard explanation of the change, but one might ask why the highest honors for scientific accomplishment continue to be accorded more to individuals than to groups. There are indeed the group awards (Watson & Crick, for example) but generally the big honors go to individuals in spite of the modern trend. Generally speaking, creativity is inhibited when done by committee and the most underestimated aspect of successful science is the importance of a creative spark in making a new discovery.

All this probably seems far afield from the question of why the science of global warming got derailed, but it could very well be that the entire edifice is built not on creative science but instead on the respectable shoulders of thoroughly ordinary scientists who – like most of us – really have little of earth-shaking import to offer. I do not mean this as criticism of those scientists since it is the fate of the entire world to be, for the most part, ordinary. But when combined with its implications for how we as humans must live our lives, the global warming theory is indeed earth-shaking and for this reason alone we should accept it only if it really is true.

There is a third explanation for the global warming mania that cannot be lightly dismissed – a problem that has frequently been discussed but that should not on such a basis be eliminated from mention here: pseudoscience has displaced religion. Throughout human history all societies have exhibited an attachment to religious ideas about how the world is created, how it is kept in order, and how humans should live in it. Only since the development of modern science have societies begun to abandon their concern with these questions, and the abandonment has been confined almost exclusively to those societies in which modern science has thrived. Many fault Darwin, but the pattern started earlier.

Only where science has attained near-divine respect have we seen large numbers of people openly admit to an atheistic point of view. God is, of course, the anti-Christ of science and rare is the individual scientist who has found a way to harmonize a belief in the divine with a commitment to science (although a few have done it). Generally speaking, the more one lives according to the principles of science the harder it is to conform to the beliefs of a religion. The evidence for this is in the much higher incidence of agnosticism and atheism among scientists than among the rest of the population.

In short, there is a fundamental incompatibility between science and religion: science requires skepticism whereas religion cannot tolerate it. For this reason, those who worship science as their chosen religion have made a fool's choice. Pure science requires a total commitment to logic, reason, and Spock-like detachment from all emotion. The only religion that might begin to tolerate such a view is Buddhism, and even in its case the vast majority of all Buddhists emotionally attach themselves to the Buddha himself (the Mahayana tradition). Emotions are learned as much as they are inherited and it is fair to say that religion – irrespective of whether it is real or fabricated, true or false – gives humans a basis upon which to organize their emotional responses to conditions in the world. Science can never do this.

But emotion, not reason, drives human action, so a society divorced from religion has no will and has no direction – an abhorrent condition for the average individual and a deadly one for any society. Under such circumstances, it would be natural to seek out some organizing principle around which appropriate emotional responses might be organized. In the modern world and in the modern countries, environmentalism has emerged as the vehicle best suited to such a purpose. Without doubt, many today – and especially those who consider themselves to be agnostics or atheists – look on the global environment as the new sacred ground.

People in the secular world have been flocking to the idea that we as humans should view ourselves as the conservators of the pristine natural world. We should protect it and defend it – nay, even sacrifice our lives to the preservation of its mystical wonders. The Green Peace warriors have become the (unheroic) crusaders of the 21st century and their behavior exemplifies the new commitment (commitment being the antithesis of science). Environmentalism is the new fallback religion.

Ironically, it is hard to distinguish from paganism, which is a worship of nature. After having "progressed" beyond that primitive state of understanding into a realm of revealed religion that replaces many gods with one, we now find ourselves returning to that earlier (more polytheistic) form of understanding. It is a lesson in humility being taught to all those who have viewed history as progressive rather than cyclical.

In any event, those who have argued from their position of disbelief that humans need religion and therefore invented it may not be so wrong after all. It pains one to admit it, but the global warming fiasco seems to confirm their unpalatable deduction. How else can one explain the widespread attachment of so many people to a theory that lacks concrete evidence of its veracity?

www.ingramcontent.com/pod-product-compliance
Lightning Source LLC
Chambersburg PA
CBHW070333190526
45169CB00005B/1866